STABILITY OF FUNCTIONAL
DIFFERENTIAL EQUATIONS

This is Volume 180 in
MATHEMATICS IN SCIENCE AND ENGINEERING
A Series of Monographs and Textbooks
Edited by WILLIAM F. AMES, *Georgia Institute of Technology*

The complete listing of books in this series is available from the Publisher upon request.

STABILITY OF FUNCTIONAL DIFFERENTIAL EQUATIONS

V. B. KOLMANOVSKII

Institute for Problems in Mechanics
Academy of Sciences of USSR
Moscow, USSR

AND

V. R. NOSOV

Institute for Problems in Mechanics
Academy of Sciences of USSR
Moscow, USSR

1986

ACADEMIC PRESS, INC.
Harcourt Brace Jovanovich, Publishers

London · Orlando · San Diego · New York
Austin · Montreal · Sydney · Tokyo · Toronto

COPYRIGHT © 1986 BY ACADEMIC PRESS INC. (LONDON) LTD.
ALL RIGHTS RESERVED.
NO PART OF THIS PUBLICATION MAY BE REPRODUCED OR
TRANSMITTED IN ANY FORM OR BY ANY MEANS, ELECTRONIC
OR MECHANICAL, INCLUDING PHOTOCOPY, RECORDING, OR
ANY INFORMATION STORAGE AND RETRIEVAL SYSTEM, WITHOUT
PERMISSION IN WRITING FROM THE PUBLISHER.

ACADEMIC PRESS INC. (LONDON) LTD.
24–28 Oval Road
LONDON NW1 7DX

United States Edition published by
ACADEMIC PRESS, INC.
Orlando, Florida 32887

British Library Cataloguing in Publication Data

Kolmanovskiĭ, V. B.
 Stability of functional differential equations.
 —(Mathematics in science and engineering,
 1. Functional differential equations
 I. Title II. Nosov, V. R. III. Series
 515.3'5   QA372

Library of Congress Cataloging in Publication Data

Kolmanovskiĭ, Vladimir Borisovich.
 Stability of functional differential equations.

 (Mathematics in science and engineering)
 Includes index.
 1. Functional differential equations—Numerical
solutions. 2. Stability. I. Nosov, V. R. II. Title.
III. Series.
QA372.K78 1985   515.3'5   85-9118
ISBN 0–12–417940–1 (hardcover) (alk. paper)
ISBN 0–12–417941–X (paperback) (alk. paper)

PRINTED IN THE UNITED STATES OF AMERICA

86 87 88 89   9 8 7 6 5 4 3 2 1

Contents

Preface — xi

Chapter 1. **Theoretical foundations of functional differential equations**

1. Time lags in engineering — 1
   1.1. Reactor (technological lag) — 1
   1.2. Rolling mill (transportation lag) — 3
   1.3. Ship stabilization (informational lag) — 4
   1.4. Manual control — 5
   1.5. Turbojet engine — 6
   1.6. Nuclear reactor — 8
   1.7. Infeed grinding model — 8
2. Aftereffects in physical, biological and other systems — 9
   2.1. Microwave oscillator — 9
   2.2. Systems with lossless transmission lines — 10
   2.3. Relativistic dynamics allowing for the finiteness of interaction speed — 13
   2.4. Population dynamics models — 13
   2.5. Model of immune response (predator–prey model) — 15
   2.6. Distribution of albumin in the bloodstream — 16
   2.7. Some economic applications — 17
3. Initial-value problem for functional differential equations — 18
   3.1. General equations with delay — 18
   3.2. RFDE with bounded delay — 19
   3.3. RFDE with unbounded delay — 21
   3.4. NFDE — 21
   3.5. Two useful lemmas — 23
   3.6. Estimations of solutions growth — 24
   3.7. Methods for solving an FDE — 25
4. Linear equations — 25
   4.1. General form of a linear RFDE — 25
   4.2. *A priori* solution estimate — 26
   4.3. Linear NFDEs — 27
   4.4. FDEs with small delays — 27
   4.5. Variation-of-constant formula — 28
5. Linear autonomous equations — 31
   5.1. Characteristic function — 31

v

	5.2.	Zeros of characteristic function	31
	5.3.	Representation of a solution as an infinite series	33
	5.4.	Linear autonomous NFDEs	33
6.		Feedback systems with aftereffects	35
	6.1.	Single-loop feedback systems	35
	6.2.	Effects of little delays	36
	6.3.	Analysis of single-loop feedback systems	38
	6.4.	Two-loop systems with delay	39

Chapter 2. Stability of retarded equations

1. General definitions of stability — 44
   - 1.1. Definition of Liapunov stability — 44
   - 1.2. Uniform Stability — 46
   - 1.3. Asymptotic stability — 46
   - 1.4. Other stability definitions. Stability under delay disturbances — 47
2. Linear autonomous retarded equations — 48
   - 2.1. Statement of the problem — 48
   - 2.2. General theorems — 50
3. Methods of stability investigation of linear autonomous RFDEs — 53
   - 3.1. Analytical methods (Pontriagin's and Chebotarev's theorems) — 53
   - 3.2. Method of D subdivision. Vyshnegradskii diagrams — 55
   - 3.3. Michailov, Nyquist and integral frequency criteria — 60
   - 3.4. Stability for arbitrary delays. Tsypkin criterion — 64
   - 3.5. General function (2.11) — 66
4. Stabilization system of the two-reflector antenna — 67
5. Liapunov direct method for equations with delay — 71
   - 5.1. Application of Liapunov functions. Razumikhin-type theorems — 72
   - 5.2. Method of Liapunov–Krasovskii functionals for equations with bounded delay — 72
   - 5.3. Equations with unbounded delay — 74
   - 5.4. Stability in the first approximation and under steady acting perturbations — 75
   - 5.5. Case of a nonpositive derivative — 76
   - 5.6. Global stability — 78
   - 5.7. Survey of other results. Exponential stability — 79
6. Construction of Liapunov–Krasovskii functionals — 80
   - 6.1. Linear autonomous equations — 80
   - 6.2. Stability of a chemical reactor closed by a P controller — 81
   - 6.3. Scalar nonlinear equations — 83
   - 6.4. Nonlinear equations of second order — 85
7. Stability of nuclear reactors — 87
   - 7.1. Single-temperature reactor with convective feedback — 87
   - 7.2. Stability of two interconnected reactors — 87
8. Mathematical models in immunology — 90
   - 8.1. Models of virus disease — 90
   - 8.2. Analysis of model (8.2), (8.3) — 91
   - 8.3. Discussion — 94

Contents vii

	9.	Design of adaptive controller for retarded systems	96
		9.1. Scalar equations	96
		9.2. Delay adjustment	100
		9.3. Multidimensional systems	101
	10.	Stability of viscoelastic bodies	103
		10.1. Constitutive equations of viscoelastic bodies	103
		10.2. Dynamic stability of a viscoelastic bar	104
	11.	Absolute stability of systems with delay	106
	12.	Stability of linear periodic equations	108
		12.1. General stability theorem	108
		12.2. Particular classes of RFDEs	108
		12.3. Floquet theory	110

Chapter 3. **Stability of neutral functional differential equations**
1. Stability of linear autonomous NFDEs   113
   1.1. General stability theorems   113
   1.2. Methods of stability investigation of linear autonomous NFDEs   116
2. Stability of aeroautoelasticity equations   120
   2.1. Equations of unsteady motion of an elastic rigid body   120
   2.2. Stability of aeroautoelastic equations   123
3. Liapunov direct method for neutral-type equations   126
   3.1. Introduction   126
   3.2. Some definitions   126
   3.3. General theorems   127
   3.4. Global stability   128
   3.5. Stability of the functional inequalities   129
   3.6. Equations with bounded delay   130
   3.7. Examples   131
4. Construction of degenerated functionals for concrete systems   132
   4.1. Stability of a chemical reactor closed by a PD controller   132
   4.2. Stability of one-dimensional nonlinear systems   134
   4.3. Use of degenerate functionals for stability investigations of RFDEs   140
5. Instability of neutral-type equations   142
   5.1. Statement of the problem   142
   5.2. Influence of the choice of an admissible class of disturbances on stability   143
   5.3. Instability conditions   144
   5.4. Connection between instability of NFDEs and ordinary differential equations   146
   5.5. Other instability conditions   147
   5.6. Instability of equations with bounded delay   148
   5.7. Distributed self-oscillatory systems   150
6. Asymptotic properties of neutral-type equations   151
   6.1. Some introductory remarks   151
   6.2. Basic definitions   152
   6.3. Stability of invariant sets   153
   6.4. Instability   154
   6.5. Interconnection between functional and differential equations   155

	7.	Liapunov functionals depending on derivatives	156
		7.1. General stability theorem	156
		7.2. Examples	158
	8.	Stability and boundedness of linear nonhomogeneous equations	159
		8.1. Connection between stability and boundedness	159
		8.2. Boundedness of derivatives	159
	9.	Linear periodic equations	160
		9.1. Some examples	160
		9.2. Stability theorem	162

Chapter 4. **Stability of stochastic functional differential equations**

1. Some prerequisites from the theory of stochastic retarded functional differential equations — 164
   1.1. Statement of the initial-value problem — 164
   1.2. Existence theorem — 165
   1.3. Itô's formula — 166
   1.4. Equations for moments — 166
2. Formulation of stability problems for SRFDEs. Liapunov direct method — 167
   2.1. Basic definitions of stability — 167
   2.2. Asymptotic p stability — 168
   2.3. Exponential p stability — 170
   2.4. SRFDEs with random delays — 170
3. Stability of scalar stochastic equations — 171
   3.1. Preliminaries — 171
   3.2. Case of autonomous linear part — 171
   3.3. Case of nonautonomous linear part — 174
4. Stability of second-order equations — 175
   4.1. General assumptions — 175
   4.2. Stability conditions — 176
   4.3. Stability domains of linear autonomous systems obtained by the Liapunov direct method — 180
5. Stationary and periodic solutions of stochastic retarded equations — 187
   5.1. Existence of stationary solutions — 187
   5.2. Relation between stability and stationarity of solutions — 187
   5.3. Periodic solutions — 189
   5.4. Ergodic properties of stationary solutions — 190
6. Stability with respect to the first approximation — 191
   6.1. General theorem — 191
   6.2. Scalar equations — 192
7. Neutral-type stochastic functional differential equations — 193
   7.1. Definition of SNFDEs — 193
   7.2. Existence theorem — 194
   7.3. Stability conditions — 194
8. Stability of linear autonomous equations — 196
   8.1. Systems of linear equations — 196

8.2.	Corollary for nth-order equations	198
8.3.	Necessary and sufficient stability conditions of scalar equations	199

References   202

Index   215

Preface

This book provides an introduction to the structure and stability properties of solutions of functional differential equations (FDEs). The purpose of this book is to introduce mathematicians, physicists, engineers and other scientists to this subject and to give some ideas about the many applications of FDEs. The beginning reader may find it useful to learn the main results, corollaries and examples from the theory and applications of FDEs, and great pains have been taken to make the text accessible to both engineers and mathematicians. Our guiding idea is clarity of exposition; thus, proofs are usually given unless precluded by excessive length or complexity, in which case the appropriate reference is cited. It is intended that the notes and references should help the reader to explore the journal literature further. The book consists of a number of more or less independent sections designed to illustrate the most important results of FDE theory and their applications. A student with a good knowledge of calculus should find all of it accessible, but the material of the book should also be appropriate for two-semester graduate-level courses.

The hereditary systems (or the systems with aftereffect, with time lag or with delay) are of great theoretical interest and form an important class as regards their applications. This class of systems is described by functional differential equations, which are also called differential equations with deviating arguments. According to Myshkis [156(3)] the *functional differential equation* is an equation involving the function $x(t)$ of one scalar argument t (called time) and its derivatives for several values of argument t.

Among functional differential equations one may distinguish some special classes of equations — *retarded functional differential equations, neutral functional differential equations, differential-difference equations, functional or difference equations,* etc. The following abbreviations are constantly used:

FDEs   functional differential equations
RFDEs   retarded functional differential equations
NFDEs   neutral functional differential equations
SFDEs   stochastic functional differential equations
SRFDEs stochastic retarded functional differential equations

If the value of the highest derivative at time t depends only on the values of lower derivatives at preceding times $t + \theta$, $-\infty < \theta \leq 0$, we get an RFDE. Retarded functional differential equations describe those systems or processes whose rate of change of state is determined by their past and present states.

If the rate of change of state depends not only on the past states but also on the past rates of change of state, then such systems are governed by an NFDE.

A functional or difference equation is an equation which involves the values of the function $x(t)$ at different times $t + \theta$, $-\infty < \theta < \infty$ or at times t_j, $j = 1, 2, \ldots$.

Historically the first FDEs were considered by Euler, Bernoulli, Lagrange, Laplace, Poisson and others in the eighteenth century [192]. The sources of FDEs for these mathematicians were various geometrical problems. In the nineteenth century and at the beginning of the twentieth century, FDEs were studied very occasionally.

The situation changed radically in the 1930s and 1940s. At that time a number of important scientific and technical problems demanded that their adequate description take into account existing different delays. The first problems of this type were considered by Volterra (viscoelasticity, 1909 [230(1)]; predator-prey models, 1928-1931 [239(2), 239(3)]), Kostyzin (problems of mathematical biology, 1934 [110]), Callender and Stevenson (instability in systems with time lag, 1936 [32]), an editorial article in *Engineer* (damping effect of time lag, 1937 [225]), Gorelik (microwave oscillator, 1939 [69]), Voznesensky and Solodovnikov (influence of hydroshock on turbine stability 1934, 1941 [243, 219(1)]), Bogomolov (feedback system for a hydroelectric power station, 1941 [25]), Sievert (reaction of cells to x-ray irradiation, 1941 [208]), Minorsky (ship stabilization, 1942 [149(2)]), Andronov and Mayer (time lag in feedback systems, 1946 [4]), Kabakov and Sokolov (control process for steam pressure, 1946 [93, 217]), Gerasimov (feedback system for heat transfer, 1949 [62]) and many others.

The appearance of such practically important problems stimulated the interest of mathematicians in investigating these not very well-known equations. In 1939 Tychonov [227] considered the functional equation of the Volterra type. In 1942 Pontryagin [175] obtained some fundamental results about the distribution of quasi-polynomial zeros. In the early 1940s Chebotarev [37] published a number of papers devoted to the Routh-Hurwitz problem for quasi-polynomials. The rapid growth of the theory of the FDE began in the 1950s. The paper of Myshkis [156(1)] has an important effect on the development of the FDE. In his paper for the first time the initial-value problem was correctly formulated. Myshkis introduced a general class of linear RFDEs and studied some basic properties of their solutions. These results have been stated in the book [156(3)], which was the first entirely devoted to the theory of the RFDE. At the same time papers by Bellman [17], Tsypkin [232(1)–232(3)], Neimark [160(1)], Leontjev [125], Wright [247(1)–247(3)] and Zubov [252(1)] appeared in which linear autonomous equations and their stability were studied in detail. Stability investigations of these equations are closely connected with study of special analytical functions representing some types of quasi-polynomials. The distribution of the quasi-polynomial zeros was investigated by several different methods — analytical (Pontriagyn, Chebotarev), frequency criteria (Kabakov, Tsypkin) and the method of D subdivision (Meiman, Neimark). The stability conditions of general linear autonomous equations with distributed bounded delay were obtained by Krasovskii with semi-group methods. It was important for the theory of equations with deviating argument that El'sgol'tz [58(3)] and Kamenskii [95(1)] distinguished different classes of the class of NFDEs.

The use of Liapunov's direct method for investigation of stability criteria of equations with delay encountered some principal difficulties [58(1)]. Krasovskii suggested the use of functionals defined on equations' trajectories instead of Liapunov functions. In particular, for equations with bounded delay, Krasovskii proved the inversion theorem and theorems about stability in the first approximation and under steady-acting perturbations [114(1)–114(6)]. Another approach to the stability problems for equations with delay based on a modification of the Liapunov function method was proposed by Razumikhin [180(1), 180(2)]. More recently, a number of new important results in stability theory were obtained — a generalization of Liapunov's direct method for NFDEs and SFDEs, stability, criteria for linear periodic equations, instability theorems, critical cases, etc.

Our first chapter is devoted to the theoretical basis of the FDE with examples from different pure and applied problems in which delays were taken into account. The second chapter deals with the stability of RFDEs. First, linear autonomous systems and corresponding quasi-polynomials are studied; then stability of feedback systems for the case of a two-reflector antenna is considered. Attention is given to Liapunov's direct method and to the construction of concrete Liapunov–Krasovskii functionals. On the basis of these functionals the stability of nuclear reactors is studied. Other problems presented concern some mathematical models of virus diseases and the design of adaptive controllers with time lag. Finally the stability conditions of viscoelastic systems and absolute stability criteria for retarded control systems are derived.

The third chapter is devoted to the stability of NFDEs. After the theoretical investigation of stability of linear autonomous equations, the interesting stability problems of unsteady motion in a continuous medium (for a model of aeroautoelasticity) are considered. Further, a particular method for stability investigation of NFDEs proposed by the authors is described. This method is based on the use of degenerated functionals. Different stability and instability theorems and examples are given. The conditions of self-exciting oscillation in distributed systems are established and finally the asymptotic behaviour of solutions of NFDEs is considered. Chapter 4 is devoted to SFDEs. Some aspects of the theory of these equations and general methods of stability investigation are formulated. Stability conditions of specific systems are given and the asymptotic properties of solutions of SFDEs are studied.

We use double enumeration of theorems and formulae inside the same chapter and triple enumeration when they are referred to in other chapters. We provide an adequate but not complete reference list. Some of the publications cited include rather extensive references to the original papers.

We would like to thank the many people to whom we are deeply indebted, particularly, M. A. Krasnoselskii, A. D. Myshkis, M. I. Monastyrskii, L. E. El'sgol'tz, R. Z. Hasminskii, A. M. Zverkin, L. E. Shaichet and D. D. Bainov for their most valuable help during the preparation of this volume.

Chapter 1

Theoretical Foundations of Functional Differential Equations

§1. TIME LAGS IN ENGINEERING

In this section we shall give some examples of engineering systems which require taking into account the systems aftereffects for their accurate qualitative and quantitative descriptions.

1.1. Reactor (Technological Lag)

Reactors and mixing tanks are standard elements of chemical engineering systems [143, 203]. Consider the system shown in Fig. 1.1. The reaction vessel (or mixing tank) is filled with two liquids through pipes A and B. There is a tap on pipe B. After a chemical reaction (or mixing), the liquid leaves the vessel through pipe C. The pH value of the liquid in pipe C is measured with a pickup. The pickup signal transformed by a controller reaches an actuator that sets the tap into motion. The physical and chemical processes in the reactor are distinguished by their complexity. Usually a simplified reactor model is used which describes the reactor by the transfer function $W_r(s) = K \exp(-sh)/(Ts + 1)$. The factor $\exp(-sh)$ is introduced in the transfer function to take into account the fact that a change in the volumetric rate of liquid entering the system causes a change in the amount of liquid leaving the system only after a time h. In reality, a certain time is needed for the mixing of liquids in the vessel, for a chemical reaction and for the transportation of liquid from pipe B to C. This time is assumed to be h. The time lag can be large enough. For example, the delay that arises in the description of the process of cellulose cooking is about 3 min, the delay in the description of the limestone drying process in rotating furnaces reaches several tens of minutes [68].

2   1. Theoretical Foundations of Functional Differential Equations

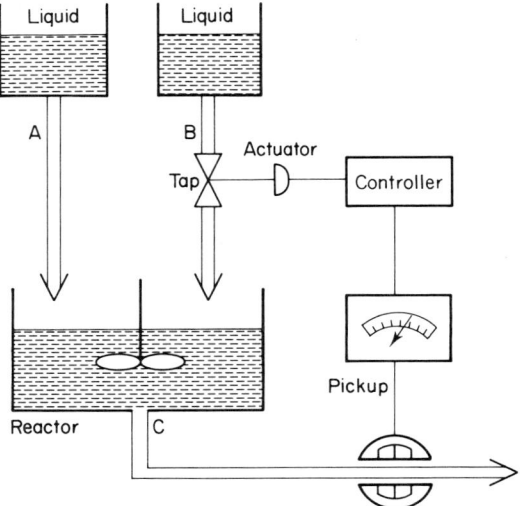

Fig. 1.1. Feedback system for a chemical reactor.

Such a time lag can be called a *technological lag*. We come across technological lags in chemical engineering (production of sulfuric acid or glass, drying, and so on [68, 203], in thermal power [62, 173], in economic models [33, 60, 102, 174], etc. The stability of reactors closed by different controllers with regard for hysteresis and saturation in an actuator is investigated in Chapter 2 §6 and Chapter 3 §4.

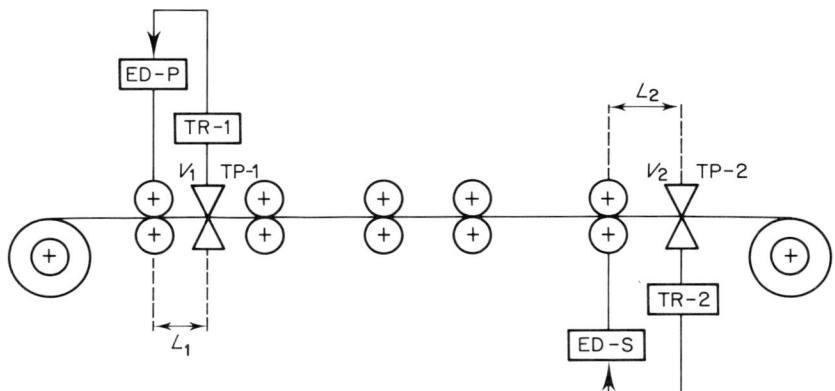

Fig. 1.2. Scheme of rolling on the continuous rolling mill.

§1. Time Lags in Engineering

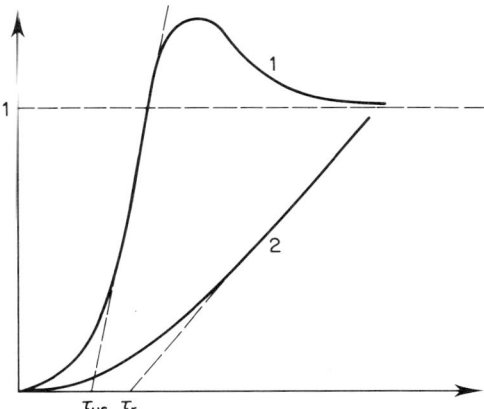

Fig. 1.3. Determination of equivalent time lag: curve 1, response on the unit step; curve 2, response on the ramp input.

1.2. Rolling Mill (Transportation Lag)

The regulation of the thickness of the outgoing plate is one of the most important problems in designing a rolling mill feedback system [56]. Figure 1.2 represents a scheme of rolling on the continuous rolling mill. In this figure the following notation is used: Rolling rates V_1, V_2 after the first and last stands change from 0.5–1.5 m/sec (strip setting-up rate) to 25–35 m/sec (operative rolling rate); TR-1, TR-2 are input and output plate thickness regulators; TP-1, TP-2 are plate thickness pickups; ED-S represents stand electric drives; ED-P means electric drives of housing screw; and L_1, L_2 are the distances from actuators ED-S and ED-P to pickups equal to ~ 1.5 m.

The adequate description of the rolling mill requires taking into account a time lag τ equal to $\tau = \tau_t + \tau_{ee} + \tau_{ep}$. Here $\tau_t = L_i/V_i$, the transportation lag which varies from 0.05 to 1 sec; τ_{ee} is the *equivalent time lag* of stand electric drive equal to 0.3 to 0.4 sec; and τ_{ep} is the equivalent time lag of pickup equal to 0.1 to 0.3 sec. Time lags τ_{ee} and τ_{ep} may be found by using transient responses. The method of their determination is illustrated by Fig. 1.3 in which the responses on the unit step and ramp inputs are shown (for details see Draljuk and Sinaiskii [56] and Rotach [189]).

Both types of lags often occur in other problems also. The *transportation lag* is usually encountered when substances, energy or signals are transmitted to certain distances. It is necessary to take this type of lag into account in heat power engineering (tranportation time of fuel from tank to furnace), in astronautics (time of wave propagation from the Earth to the Moon) and so on.

1.3. Ship Stabilization (Informational Lag)

Consider the simplest model of ship stabilization [160(2)]. Assume that ship dynamics is described by

$$I\ddot{\varphi} + h\dot{\varphi} = -K\psi, \quad I > 0, \quad K > 0, \quad h \gtrless 0 \tag{1.1}$$

where φ is the ship deviation angle and ψ the turning angle of the rudder. Assume also that the change of rudder angle ψ is governed by the automatic helmsman rule

$$T\dot{\psi} + \psi = \alpha\xi + \beta\dot{\xi}, \quad T > 0 \tag{1.2}$$

Here ξ is a measured value of the ship deviation angle. In practice $\xi(t) \neq \varphi(t)$ because it is impossible to measure ship deviation instantaneously. Assume that

$$\xi(t) = \varphi(t - \tau) \tag{1.3}$$

Under these hypotheses the ship with an automatic helmsman is described by an RFDE

$$TI\dddot{\varphi}(t) + (Th - I)\ddot{\varphi}(t) - h\dot{\varphi}(t) + K\beta\dot{\varphi}(t - \tau) + K\alpha\varphi(t - \tau) = 0 \tag{1.4}$$

One of the important problems is the choice of automatic helmsman parameters α and β which guarantee the stability of a closed system (1.4). Notice that the measure rule (1.3) is often replaced by

$$\tau\dot{\xi}(t) + \xi(t) = \varphi(t) \tag{1.5}$$

Such replacement sometimes yields true qualitative conclusions. In Figure 1.4 the full line limits the stability domain for the measure rule (1.3) and the dotted line for rule (1.5) [160(2)].

The time lag arising in this problem can be called an *informational lag*. It may occur in a number of control systems which contain subsystems of

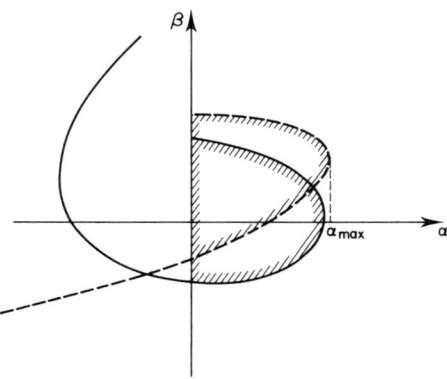

Fig. 1.4. Stability domains: solid line, measure rule (1.3); dashed line, measure rule (1.5).

§1. Time Lags in Engineering

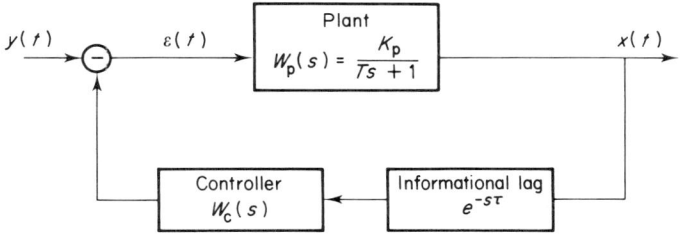

Fig. 1.5. Control system with informational lag.

transfer and processing of information. An informational lag is usually modelled by a unit with transfer function $\exp(-s\tau)$ (see Fig. 1.5).

1.4. Manual Control

In man–machine systems it is important to have a model of human response [22, 24, 201]. Figure 1.6 shows a structural scheme of manual control for a single-loop system. Here $W_r(s)$, $W_p(s)$ and $W_m(s)$ are transfer matrices of the signal reproduction system, of the plant and of the measure system, respectively, and n_i are external noises. Man in this scheme is represented by a nonlinear unit with input $y(t)$ and output $c(t)$. Man perceives input signals by image, sound and acceleration channels. Man's output signal is a muscle response. Experimental researches [24, 201] shows that in first approximation man's transfer function is given by

$$W_h(s) = \frac{K \exp(-s\tau)}{1 + sT_1} \frac{1 + asT_1}{1 + sT_2} \qquad (1.6)$$

Here lag τ evaluates the time interval necessary for the signal to reach the human central nervous system and then the muscles. This lag lies in the interval 0.15 to 0.25 sec, and it is almost constant for every man. Time

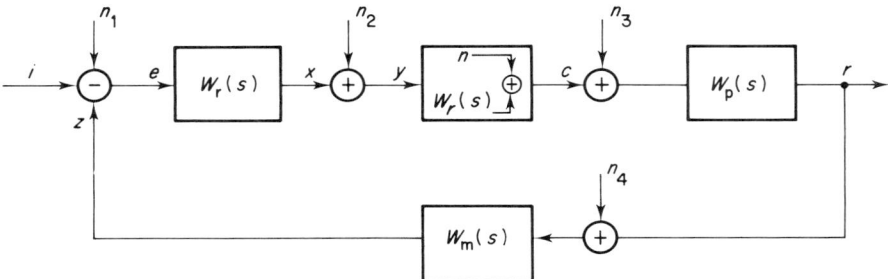

Fig. 1.6. Stochastic model of manual control.

Table 1.1

Signal frequency	T_1	T_2	τ	a	K
0.16	25	0.67	0.15	0.08	100
0.32	9.1	0.2	0.20	0.055	40
0.48	5	0.09	0.25	0.067	15

constant T_1 characterizes the limitation of muscle reaction speed and approximately equals 10 sec. The factor a takes into account the degree of instruction, experience of the operator or type of problem. The process of instruction is more prolonged than is the process of control. For this reason it is assumed that factor a in (1.6) is a constant. Table 1.1 [24] gives experimental average values of parameters for transfer function (1.6) for different frequencies of input signal.

More adequate than (1.6) is a model which contains a noise $n(t)$ that does not correlate with the input signal (see Fig. 1.6). Such a system is usually described by an SRFDE (see Chapter 4).

A nonlinear model of manual control is represented in Fig. 1.7 [24]. This model allows for a dead zone of operation and dynamics of the central nervous system and muscle response.

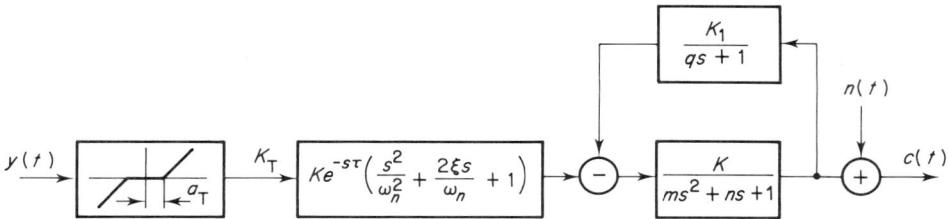

Fig. 1.7. Nonlinear model of manual control.

1.5. Turbojet Engine

Modern turbojet engines have to operate over a large range of speeds and altitudes. The main purpose of the turbojet engine is to provide speed adequate for flight. The turbojet engine contains the following basic elements: a variable-geometry supersonic air intake, a turbocompressor with principal combustion chamber, an afterburner and a propulsive nozzle [23, 41, 136, 159, 202]. In the turbojet engine, control factors include the following: G_f (the fuel quantity entering the principal combustion chamber), F_n (the propulsive nozzle section area), and F_{oc} (the cross-sectional area of the over-cross flap in air intake). Controlled variables are n (the turbocom-

§1. Time Lags in Engineering

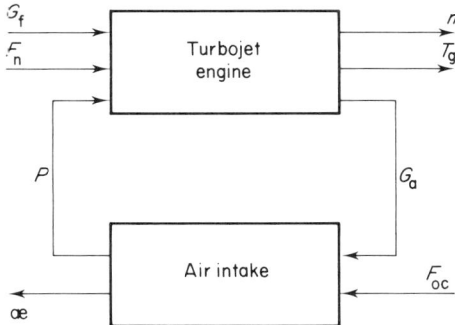

Fig. 1.8. Structural scheme of turbojet engine with air intake.

pressor rotation frequency), T_g (the gas temperature before the turbine), and æ (the intradiffuser parameter); æ $= P_æ/P_i$, where $P_æ$ is static pressure after the shock wave and P_i is impact pressure in the air intake. Variables P (inlet engine pressure) and G_a (air quantity in combustion chamber) reflect the interaction between engine and air intake (see Fig. 1.8). Linearized equations of the turbojet engine are [23, 136, 159, 202]

$$T_e\dot{x}_1(t) + x_1(t) = -K_{15}x_5(t) + K_{16}u_6(t) + K_{17}u_7(t)$$
$$T_e\dot{x}_2(t) + x_2(t) = K_{26}(M_{26}\dot{u}_6(t) + u_6(t)) - K_{27}u_7(t)$$
$$T_e\dot{x}_3(t) + x_3(t) = -K_{35}(-M_{35}\dot{x}_5(t) + x_5(t))$$
$$\qquad\qquad\qquad + K_{36}(-M_{36}\dot{u}_6(t) + u_6(t)) \qquad (1.7)$$
$$T_i\dot{x}_4(t) + x_4(t) = K_{43}x_3(t - \tau_1) + K_{48}u_8(t - \tau_1)$$
$$T_i\dot{x}_5(t) + x_5(t) = K_{53}x_3(t - \tau_2) + K_{58}u_8(t - \tau_2)$$

Here x_1-x_5 and u_6-u_8 denote increments of n, T_g, G_a, æ, P, G_f, F_n and F_{oc}, respectively (for example, $x_1(t) = (n(t) - n_{st})/n_{st}$). Also, K_{ij} and M_{ij} denote different positive constants and τ_1 and τ_2 represent time lag in intake versus flow and time lag along flow, respectively. Further, T_e and T_i are time constants of the engine and the intake. It is not possible to neglect τ_1 and τ_2 in (1.7) because $\tau_1/T_i \cong 1$ and $\tau_2/T_i \cong 1$. Sometimes one uses the simplified model in which the intake is described by the transfer function

$$W_i(s) = K_{48}[\exp(-\tau_1 s)]/(T_i s + 1) \qquad (1.8)$$

Here $\tau_1 = 1/a(1 - M)$, where 1 is the distance from the over-cross flap to the pressure pickup $P_æ$; a is a sound velocity and M is a Mach number in the subsonic part of the intake.

1.6. Nuclear Reactor

Nuclear reactors are complex feedback systems, described by different models [67, 70]. Some of these models use RFDEs. For example, in [70] the following model is investigated:

$$\dot{N}(t) = [-\alpha T(t - h) - \varepsilon(N(t) - N_{st})]N(t)$$
$$\dot{T}(t) = r[N(t) - N_{st} - T(t)], \quad \alpha, \varepsilon\, r > 0$$

Here $N(t)$ denotes reactor power, N_{st} is a stationary reactor power, $T(t)$ is a deviation of reactor temperature from the stationary value and h is a time lag. One of the important problems of nuclear reactor analysis is a study of their stability under different disturbances. Some stability results are given in Chapter 2 §7 (see also Gorjachenko [70]).

1.7. Infeed Grinding Model*

A mathematical model of grinding has to take into account the action of grinder element masses (m_1, m_2), dissipative forces (F_ξ), elastic forces (F_c), stiffness of the piece and grinding wheel contact and other phenomena. The grinding process is schematically shown in Fig. 1.9. As control factors we take the cross-feed rate V and the angular velocity of the piece rotation ω_1 (round-feed). Output variables are the radial F_y and tangential F_z components of the cutting effort. In the linear model of grinding one uses the transfer functions

$$W_1(s) = F_y(s)/V(s) = D^{-1}(s)\left[\sum_{i=0}^{3} c_i s^i + \sum_{i=0}^{3} d_i s^i \exp(-\tau_1 s)\right]$$

$$W_2(s) = \frac{F_y(s)}{\Delta\omega(s)} = D^{-1}(s)\left[\sum_{i=0}^{5} h_i s^i + \sum_{i=0}^{4} g_i s^i \exp(-\tau_2 s)\right] \quad (1.9)$$

$$D(s) = \sum_{i=0}^{5} a_i s^i + \sum_{i=0}^{5} b_i s^i \exp(-\tau_1 s)$$

Here $\tau_1 = 2\pi\omega^{-1}$, $\tau_2 = 2\pi(\omega + \Delta\omega)^{-1}$ and $a_i, b_i, c_i, d_i, h_i, g_i$ are constants. The time lag appearance in this model is conditioned by the dependence of the grinding process on the surface state at the previous rotation. Relations (1.9) show that grinding can be described by an NFDE of fifth order. Other cutting models using RFDEs have been studied by Shil'man [204].

---

* From Michelkevich and Chabanov [146].

§2. Aftereffects In Physical, Biological and Other Systems

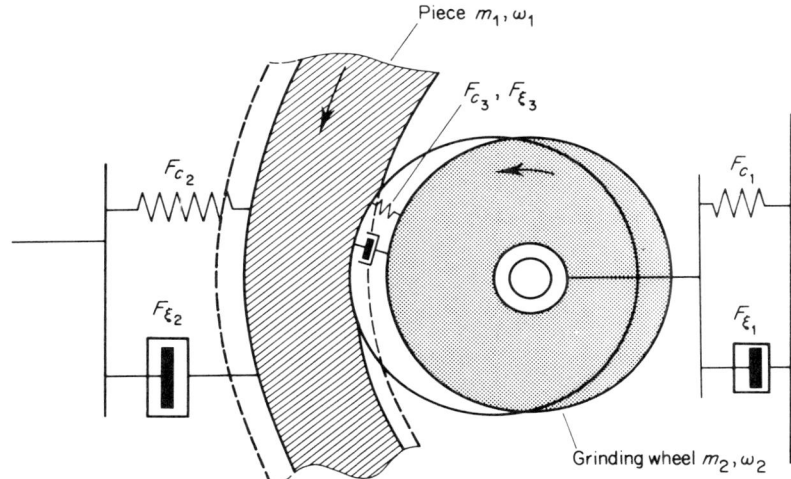

Fig. 1.9. Scheme of infeed grinding.

§2. AFTEREFFECTS IN PHYSICAL, BIOLOGICAL AND OTHER SYSTEMS

2.1. Microwave Oscillator

According to Gorelik [69] the equation

$$\ddot{x}(t) + 2r\dot{x}(t) + \omega^2 x(t) + 2qx(t-1) = \varepsilon x^3(t-1) \qquad (2.1)$$

describes the vacuum-tube oscillator (Fig. 1.10) and takes into account the time lag caused by the finiteness of the time that the electrons require to pass

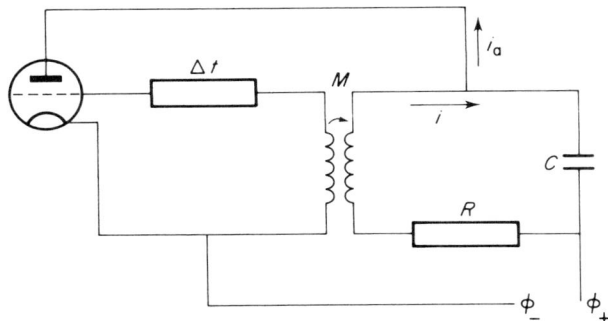

Fig. 1.10. Microwave vacuum-tube oscillator.

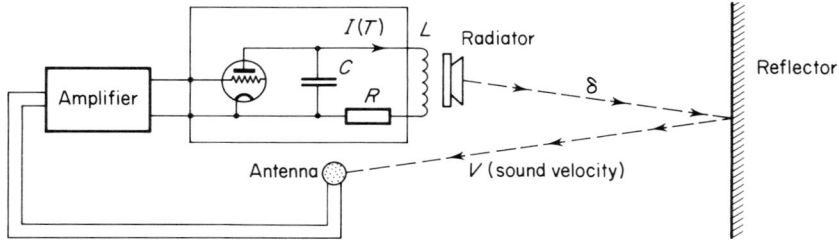

Fig. 1.11. Radio system with transportation time lag.

from the cathode to the anode of a vacuum tube. For a similar problem in Rubanik [193] an equation of the Van der Pol type (2.2) is used:

$$\ddot{x}(t) + \alpha x(t) - f(x(t-\tau))\dot{x}(t-\tau) + x(t) = 0 \qquad (2.2)$$

The greatest interest in investigation of Eqs. (2.1) and (2.2) is the search for periodic solutions and their stability studies [9, 66].

An equation similar to (2.1) has been obtained by Minorsky [149(2)]. Minorsky dealt with the system shown on Fig. 1.11. For current $I(t)$ he has obtained

$$LC\ddot{I}(t) + RC\dot{I}(t) + I(t) + AKL\dot{I}\left(t - \frac{2\delta}{\sigma}\right) = BK^3L\dot{I}^3(t - 2\delta/\sigma)$$

The interaction of coupled oscillators is described by Starik [221] as

$$\ddot{y}(t) + [\omega^2 + \varepsilon\lambda \sin \varphi(t-\tau_1)]y(t) = -\varepsilon[h\dot{y}(t) + \gamma y^3(t-\tau_2)]$$
$$I\ddot{\varphi}(t) = \varepsilon[L(\dot{\varphi}(t)) - H\varphi(t) - \sigma_1 y^2(t-\tau_3)\cos \varphi(t)$$
$$- \sigma_2 \sin \varphi(t) - \sigma_3 \cos \varphi(t)]$$

The feedback system stability for the two-mirror spherical antenna is investigated in Chapter 2 §4.

2.2. Systems with Lossless Transmission Lines

Neutral functional differential equations are frequently used for the study of distributed networks containing lossless transmission lines. In Bogomolov [25] and Solodovnikov [219(1)] NFDEs appeared in studies of feedback systems for the hydraulic turbine. Let us consider in detail one example of this type. Let the system consist of a long electrical line (cable) of length l, one end of which is switched on a power source E with resistance R, while the other end is switched on an oscillating circuit formed by a condenser C_1 and a nonlinear element, the volt–ampere characteristic of which is $i = g(v)$ (Fig.

§2. Aftereffects In Physical, Biological and Other Systems

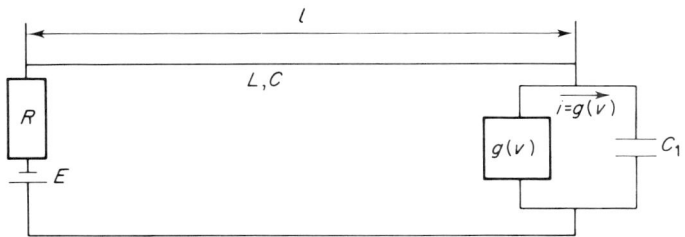

Fig. 1.12. System with lossless transmission line.

1.12). Let L and C denote the linear inductance and capacitance of a long line, respectively, and assume that the line is lossless. The processes in such a system are described by the hyperbolic partial differential equations [28, 108(8), 147, 148]

$$Li_t(x, t) = -u_x(x, t), \qquad Cv_t(x, t) = -i_x(x, t)$$
$$0 < x < l, \qquad 0 < t \qquad (2.3)$$

with boundary conditions

$$E - v(0, t) - Ri(0, t) = 0, \qquad C_1 v_t(l, t) = i(l, t) - g(v(l, t)) \qquad (2.4)$$

Let $s = (LC)^{-1/2}$ denote the wave impedance of the long line. It is known that the general solution of Eq. (2.3) can be represented as

$$V(x, t) = \varphi(x - st) + \psi(x + st)$$
$$i(x, t) = Z^{-1}[\varphi(x - st) + \psi(x + st)]$$

where φ and ψ are arbitrary smooth functions. From these expressions it follows that

$$2\varphi(-st) = v[l, t + (l/s)] + Zi[l, t + (l/s)]$$
$$2\psi(st) = v[l, t - (l/s)] - Zi[l, t - (l/s)]$$

From the first boundary condition (2.4) at the moment $(t - l/s)$, we have

$$i(1, t) - Ki[l, t - (2l/s)] = \alpha - Z^{-1}v(l, t) - KZ^{-1}v[l, t - (2l/s)]$$
$$K = (Z - R)(Z + R)^{-1}, \qquad |K| < 1, \qquad \alpha = 2E(Z + R)^{-1}$$

From the second boundary condition (2.4) designating $x(t) = v(l, t)$, we obtain

$$(1/C_1)(d/dt)[x(t) - Kx(t - \tau)]$$
$$= \alpha - Z^{-1}x(t) - KZ^{-1}x(t - \tau) - g(x(t)) + Kg(x(t - \tau))$$
$$\tau = 2l/s \qquad (2.5)$$

Equation (2.5) is a nonlinear NFDE. Thus, the investigation of linear partial derivative system (2.3) with nonlinear boundary conditions (2.4) is reduced to the study of a nonlinear NFDE. This reduction may also be used in other problems with distributed lines [104, 148].

In Danilushkin and Rapoport [51] the process of *continuous induction heating of a thin moving body* is investigated. This process can be described by the equation

$$\theta_t(t, x) + V(t)\theta_x(t, x) - (c\gamma)^{-1}F(t, x) + \beta(t, x)\theta(t, x) = 0$$

with boundary conditions

$$\theta(0, x) = \theta_0(x), \qquad \theta(t, 0) = 0, \qquad 0 \leq t, \quad x < \infty$$

Here $\theta(t, x)$ is the temperature of the heating body at moment t in point x; $V(t)$ is body speed; α, c, γ, β are several thermal coefficients; and $F(t, x)$ is specific heat release. The control factors in this system are $V(t)$ and $F(t, x)$. Using the structural method developed in Butkovskii [31], one can obtain transfer functions represented in Fig. 1.13. Closing the system, we obtain, depending on the type of controller, an RFDE or an NFDE as an equation describing the process.

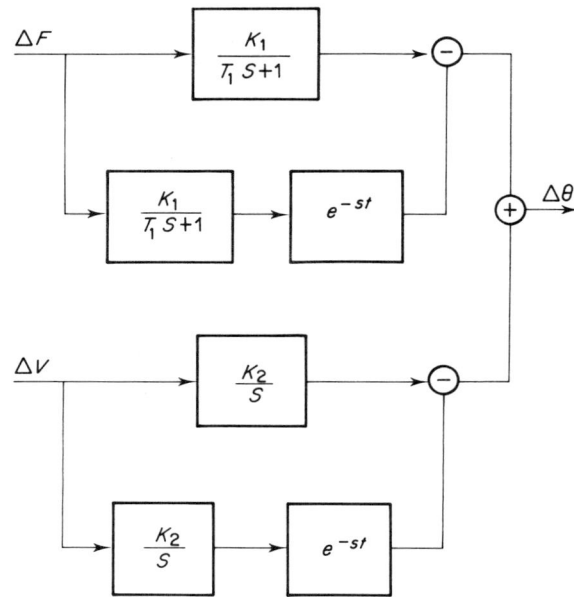

Fig. 1.13. Structural scheme of continuous induction heating.

2.3. Relativistic Dynamics Allowing for the Finiteness of Interaction Speed

Driver [57(1)] has proposed taking into account the finiteness of interaction speed in the description of relativistic particles. According to Zhdanov [251], the *equations of motion of two interacting particles* are

$$\dot{V}_i(t) = f_{ij}(x_t), \qquad \dot{r}_i(t) = V_i(t), \qquad (i,j) = (1,2)(2,1)$$

$$f_{ij}(x) = \int_{-\infty}^{0} ds\, \theta(-s) K_{ij}$$

$$\times [\dot{V}_j(t+s), V_i(t), V_j(t+s), r_i(t) - r_j(t+s)]$$

$$\times \delta[s^2 - |r_i(t) - r_j(t+s)|^2]$$

$$K_{ij}[w, u, v, r] = K_{ij}^{(1)}[u, v, r] + K_{ij}^{(2)}[w, u, v, r]$$

Here the functions $K_{ij}^{(1)}$ and $K_{ij}^{(2)}$ are homogeneous in r of orders -2 and -1, respectively, and are algebraic in u and v. The function $K_{ij}^{(2)}$ is linear in w and $m_i K_{ij}^{(1)}[0,0,r] = \ae r|r|^{-3}$. Note that the typical equations of relativistic dynamics are NFDEs with unbounded delay, depending on the solution [57(1), 171(1), 171(2), 251].

In a recent paper [252(3)] Zubov has considered the *problem of relativistic particle motion in a central field*. The equation of motion of this particle is $m\ddot{r}(t) = -kr(t - \tau|r|)|r(t - \tau|r|)|^{-3}$. Here r is a vector joining the particle with the immovable centre. It is known that in such a system without delay (i.e., for $\tau|r| \equiv 0$) a unique circular orbit passes across every point (x_0, y_0, z_0) of configuration space. If we allow for the interaction delay, then the situation changes qualitatively. The circular orbits settle on the spheres if and only if their radii verify the quantization conditions.

$$\omega(|r|)\tau(|r|) = 2\pi n, \qquad n = 0, \pm 1, \pm 2, \ldots \tag{2.6}$$

Conditions (2.6) coincide with *Bohr quantization rules*. The particle energy is also quantized in this case. It is easy to calculate that an atom of dimension 10^{-8} cm having a nucleus of 10^{-12} cm contains approximately 10^2 such quantized orbits.

2.4. Population Dynamics Models

Let $N(t)$ denote the number of individuals in an isolated population at time t (or the biomass of the population). The life-span of every individual is assumed to be a fixed constant L. Assume that the number of births per unit time is some function of $N(t)$, say $g[N(t)]$. Then the number of deaths at time t is $g[N(t-L)]$. Under these assumptions the growth of the population is governed by the equation $\dot N(t) = g(N(t)) - g[N(t-L)]$ [45]. Another population model allows a distribution of life-spans. Let $P(a)$ be the probability of survival to age (at least) a and L be a maximum life-span, $P(L) = 0$. Then [45]

$$\dot N(t) = g(N(t)) + \int_0^L g(N(t-s))\dot P(s)\,ds \qquad (2.7)$$

Another known *single species growth model* has been developed by Hutchinson [88]

$$\dot N(t) = [\alpha - \gamma N(t-L)]N(t), \qquad \alpha, \gamma > 0 \qquad (2.8)$$

Equation (2.8) has been studied in several papers [104, 224]. In particular, the following results have been obtained.

(1) If $\alpha L < \frac{37}{24}$, then the stationary solution $N(t) = \alpha\gamma^{-1}$ is globally asymptotically stable with respect to any admissible disturbances $N(0) > 0$.

(2) If $\frac{37}{24} < \alpha L < \pi/2$, then the stationary solution is asymptotically stable, but not globally.

(3) If $e^{-1} < \alpha L < \pi/2$, then the solutions tend toward the stationary solution crossing it infinitely many times.

(4) If $\alpha L > \pi/2$, then there exists a periodic solution not equal to a constant.

In this last case the population is said to be *ecologically stable*, since it can exist for an infinitely long time. Kolesov has considered Eq. (2.8) in a number of papers [104, 250]. In particular, he has obtained the asymptotic formulae solutions of Eq. (2.8) for $\alpha L - \pi/2 \ll 1$ and for $\alpha L \gg \pi/2$. A more complicated population model, stated by Kolesov, is

$$\dot x(t) = r\{1 - k^{-1}[\alpha_1 x(t-h) + \alpha_2 x(t-h-1)]\}x(t), \qquad \alpha_1 + \alpha_2 = 1$$

It is shown [250] that for $h \cong 1$ a small $\alpha_2 > 0$ (existence of survivals reaching longevity) implies a decrease of oscillations, but for $h \cong 2$ any number of long-livers implies an increase of oscillations. Thus it permits us to explain the reasons for more intensive oscillations in Iakoutia (a region in the north of the USSR) of hare quantity in comparison with nutria quantity. Other generalizations of Eq. (2.8) are available [10, 40, 103, 224, 239(3)] as are other ecological models [10, 14(2), 50, 71, 246].

2.5. Model of Immune Response (Predator-Prey Model)

The following model of immune response is developed in [55]. Let $g(t)$ denote the number of antigens penetrating an organism and $a(t)$ the number of antibodies. Assume that the rate of antibody production is proportional to $g(t - \tau)$. Here τ is a mean time for organism reaction on the antigens. For real diseases τ is great enough (3-4 days). In linear approximation the dynamics of the immune process is described by

$$\dot{g}(t) = Kg(t) - Qg(t)a(t)$$
$$\dot{a}(t) = Ag(t - \tau) - Rg(t)a(t) - Ea(t) \qquad (2.9)$$

System (2.9) has two stationary points: $\text{I} = (0, 0)$ and $\text{II} = [K/Q, KE/(CAQ - KR)]$. Point I is always unstable. The stability of point II depends essentially on the delay τ. For small τ point II is asymptotically stable. This means biologically that the disease does not progress, the number of antigens is practically constant and the organism becomes an infection carrier. For great τ there is a periodic solution in System (2.9) which can be interpreted as a chronic disease. The numerical simulations show that the oscillation amplitude decreases with the decrease of τ. The time interval when the number of antigens is nearly zero (from $7 \cdot 10^{-14}$ to $8.5 \cdot 10^{-4}$ of the stationary value) decreases and reaches 0.5 to 0.7 of the oscillation period. Such an interval corresponds to almost complete convalescence, but there always exists a danger of relapse.

More complicated models of immune response have been developed by Marchuk [21, 139(1)-139(3)]. In Chapter 2 §8 we shall investigate them in detail. Other immunological models are also available [10, 14(2), 26, 27, 135].

Equations similar to (2.9) have been considered by Volterra [239(3)], Wangersky and Cunningham [245] and others [50, 104]. One paper [245] deals with the equations

$$\dot{x}(t) = ax(t) - cx(t)y(t) - bx^2(t)$$
$$\dot{y}(t) = -ey(t) + c_1 x(t - h)y(t - h)$$

and [239(3)] with the equations

$$\dot{x}(t) = b_1 x(t)\left[1 - c_{11}x(t) - c_{12}\int_0^\infty y(t-s)\,d\alpha_1(s)\right]$$
$$\dot{y}(t) = b_2 y(t)\left[1 + c_{21}\int_0^\infty x(t-s)\,d\alpha_2(s)\right]$$
$$b_i > 0, \quad c_{ij} > 0, \quad d\alpha_i(s) > 0, \quad \int_0^\infty d\alpha_i(s) = 1, \quad i = 1, 2$$

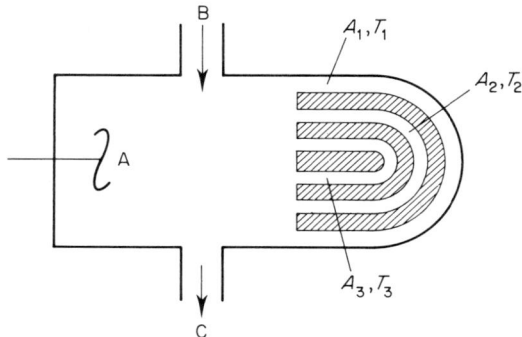

Fig. 1.14. Mechanical model of the bloodstream.

The modification of the Lotka–Volterra model has been proposed [10, 104]

$$\dot{x}(t) = r_1[1 + \alpha(1 - y(t)/K_2) - x(t - h_1)/K_2]x(t)$$
$$\dot{y}(t) = r_2[x(t)/K_1 - y(t - h_2)/K_2]y(t)$$

2.6. Distribution of Albumin in the Bloodstream*

The following mechanical model can be used to explain the distribution of albumin in the bloodstream. Consider the system shown in Figure 1.14 in which A is a unity volume tank filled with water which circulates through three pipes with volumetric flow rates A_1, A_2 and A_3 and with flow times T_1, T_2 and T_3. Assume that at $t = 0$; 1 g of dye is introduced into tank A, which is mixed instantaneously. Thus, the initial dye concentration in the tank is unity and the initial dye concentration in the pipes is zero. Dyed water leaves the system through a pipe C with volumetric rate L, and pure water enters the system at B with the same flow rate. It is assumed that no mixing takes place in the pipes, that the diffusion is neglected and that the flow velocity is assumed to be constant over a cross section of any pipe. Let $x(t)$ denote the dye concentration in the tank at time t. A material balance on dye in the tank leads to the RFDE

$$\dot{x}(t) = -(A_1 + A_2 + A_3 + L)x(t) + A_1 x(t - T_1) \\ + A_2 x(t - T_2) + A_3 x(t - T_3)$$

The initial conditions describing the dye concentration are given by $x(t) = 0$, $-T_3 \leq t < 0$, $x(0) = 1$.

Other biomedical models have been proposed [14(2), 139(3)]. To describe

---

* From Bailey and Reeve [11].

§2. Aftereffects In Physical, Biological and Other Systems

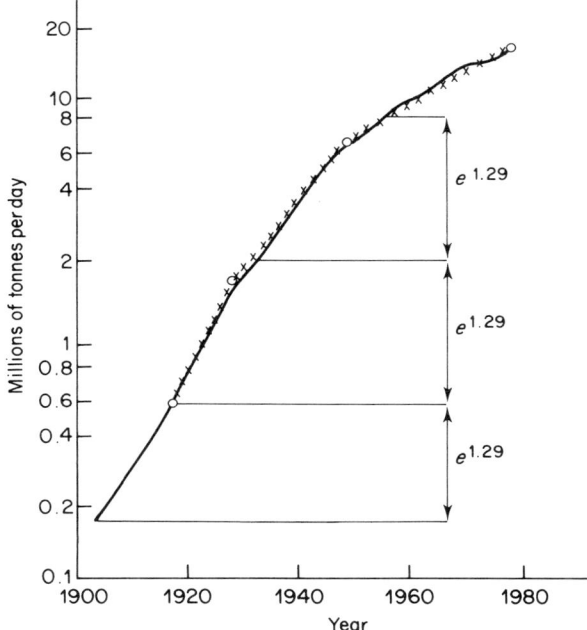

Fig. 1.15. Total petrol extraction in the USA (in millions of tonnes per day): solid line, real extraction; crossmarks, prognosis based on dependence of lag on the time of extraction.

the behaviour of the central nervous system in the process of study, Shimbell [206] has proposed the following model:

$$\dot{X}(t) = k[x(t) - x(t-1)][N - x(t)], \qquad t > 1$$
$$\dot{x}(t) = kx(t)[N - x(t)], \qquad 0 \le t \le 1, \quad x(0) = x_0 \qquad (2.10)$$

The stability of Eq. (2.10) is investigated in detail in Chapter 3 §5.

2.7. Some Economic Applications

Retarded functional differential equations are used in Kobrinskii and Kus'min [102] for description of different economic processes. The equation $\dot{y}(t) = ky[t - \tau(t)]$ allows the influence of production time on characteristics of the process and is called a *reproduction equation*. The delay $\tau(t)$ in the reproduction equation changes with the age or size of the system. The reproduction equation can be used for analysis and prognosis of economic characteristics. As examples, two prognoses are shown in Figs. 1.15 and 1.16 [102]. For an economic interpretation of Eq. (2.7), see Kobrinskii and

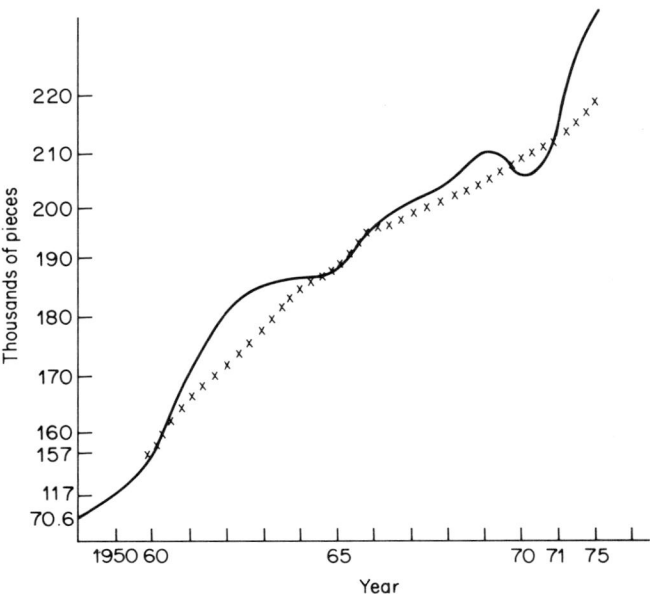

Fig. 1.16. Production of machine tools in USSR (in thousands of pieces): solid line, real production; crossmarks, prognosis based on reproduction equation.

Kus'min [102]. Other economic models which take into account the aftereffects may also be found [33, 60, 102, 174].

§3. INITIAL-VALUE PROBLEM FOR FUNCTIONAL DIFFERENTIAL EQUATIONS

3.1. General Equations with Delay

In this section we study equations with unknown functions $x(t) \in R_n$ depending on a scalar argument t. Here R_n is any real n-dimensional normed vector space with norm $|x(t)|$. Let M denote a complete metric space of continuous functions $\varphi(\theta)$ mapping the interval $(-\infty, 0]$ into R_n with the metric ρ. As M we can take the space $C[-\infty, 0]$ of continuous functions $\varphi: (-\infty, 0] \to R_n$ with a countable system of seminorm $\|\varphi\|_j$ and a metric ρ:

$$\|\varphi\|_j = \max|\varphi(t)|, \quad t \in [-j, 0], \quad j = 1, 2, \ldots$$

$$\rho(\varphi, \psi) = \sum_{j=1}^{\infty} 2^{-j} \|\varphi - \psi\|_j \cdot (1 + \|\varphi - \psi\|_j)^{-1}$$

§3. Initial-Value Problem for Functional Differential Equations

Another example of M is a normed space $CB[-\infty, 0]$ of continuous and bounded functions on $(-\infty, 0]$ with norm $\|\varphi\|_B = \sup|\varphi(\theta)|, -\infty < \theta \leq 0$. An important example of M is a space $C[-h, 0]$ of continuous functions $\varphi(\theta)$ mapping the interval $[-h, 0]$ into R_n with the usual norm $\|\varphi\| = \max|\varphi(\theta)|, -h \leq \theta \leq 0$. A number of space M is given by normed spaces, coordinated with the equations (see Chapter 2 §2 and Kolmanovskii [108(3)]), and also by spaces "with memory" [42-44, 47].

Designate Q_H a sphere in the space M

$$Q_H = \{\varphi \in M, \rho(\varphi, 0) \leq H\} \tag{3.1}$$

Denote $x_t = x_t(\theta) = x(t + \theta)$ an element of M. We say that the relation

$$\dot{x}(t) = f(t, x_t), \qquad x_t = x(t + \theta) \tag{3.2}$$

is a *general retarded functional differential equation* on M and denote it by an RFDE. In (3.2) $f: R_1 \times M \to R_n$ is a given function (map) and the overdot represents the right-hand derivative. If $-\infty < \theta \leq 0$, then Eq. (3.2) is an *RFDE with unbounded delay*. If $-\infty < -h \leq \theta \leq 0$, then (3.2) is an *RFDE with bounded delay*.

Let t_0 be an *initial moment* and let $\varphi(\theta) \in M$ be a given function. The *initial-value problem* (or Cauchy problem) for Eq. (3.2) is as follows. Define a function $x(t)$ which satisfies Eq. (3.2) for $t \geq t_0$ and for $t \leq t_0$ and which has the *initial data*

$$x_{t_0} = x(t_0 + \theta) = \varphi(\theta) \tag{3.3}$$

3.2. RFDE with Bounded Delay

The simplest RFDE is

$$\dot{x}(t) = f[t, x(t), x(t - h)], \qquad h > 0 \tag{3.4}$$

Equation (3.4) is also called a differential difference equation. The solution $x(t)$ of the initial-value problem (3.4), (3.3) can be obtained by a *step method*. This method permits us to define a continuous solution $x(t)$ of problem (3.4), (3.3) by successive integration of ordinary differential equations. Replacing $x(t - h)$ for $t \in [t_0, t_0 + h]$ in (3.4) by $\varphi(t - h)$, we obtain the ordinary differential equation

$$\dot{x}(t) = f[t, x(t), \varphi(t - h)], \qquad t_0 \leq t \leq t_0 + h$$

Having defined the solution $x(t)$ on the interval $[t_0, t_0 + h]$, we can continue in a similar way and define $x(t)$ on $[t_0 + h, t_0 + 2h]$, $[t_0 + 2h, t_0 + 3h]$ and so on. For example, the solution of the equation $\dot{x}(t) = ax(t-1)$, $t \geq 0$, with the initial condition $x_0(\theta) = 1$, $t \leq 0$, is given by the formula

$$x(t) = \sum_{n=0}^{N} a^n [t - (n-1)h]^n (n!)^{-1}, \qquad N = [th^{-1}] + 1.$$

Here $[s]$ denotes the greatest integer in s. The step method is *not universal*. For example, it cannot be applied to the equations with time-varying delays which vanish in some points:

$$\dot{x}(t) = f[t, x(t), x(t - h(t))], \qquad h(t) \geq 0, \quad h(t_i) = 0, \quad t_i > t_0$$

But the step method can be used for general equations (3.2) if $f(t, x_t)$ depends only on $x(t)$ and $x(\theta)$ for $\theta \leq \varepsilon < 0$. In some cases the step method allows us to obtain different theorems of existence, uniqueness and smoothness of solutions and other qualitative properties.

Unlike ordinary differential equations, the solution of (3.4), (3.3) even for infinitely differentiable functions φ and f has, in general, a discontinuity at the point $t = t_0$. Really, $\dot{x}(t_0 + 0) = f[0, \varphi(\theta)]$ and is not always equal to $\dot{x}(t_0 - 0) = \dot{\varphi}(0)$. At point $t = t_0 + h$ the solution has, in general, a discontinuity in the second derivative, but its first derivative is by all means continuous, and so on. Thus the solution becomes more smooth from step to step. This process is called a *smoothing of solutions*.

Let us outline another characteristic peculiarity of RFDEs. For the RFDE (3.4), a solution exists to the right of the initial point t_0 and does not exist, in general, to the left of t_0. Consider, for example, the equation

$$\dot{x}(t) = x(t-1), \qquad t \geq 1; \qquad x(t) = \varphi(t), \qquad 0 \leq t \leq 1 \qquad (3.5)$$

Rewritten in the form $x(t) = \dot{x}(t+1)$, this equation means that on $[-1, 0]$, $x(t) = \dot{\varphi}(t+1)$; on $[-2, -1]$, $x(t) = \dot{x}(t+1) = \ddot{\varphi}(t+2)$; and so on. Solution $x(t)$ is expressed by $\varphi^{(m)}(t)$ and $m \to \infty$ when $t \to -\infty$. If $\varphi(t)$ is only continuous, then the solution $x(t)$ does not exist to the left of $t = 0$. If there exists $\varphi^{(m)}(t)$ but $\varphi^{(m+1)}(t)$ does not exist, then the solution $x(t)$ of (3.5) exists only on $[-m, 0]$. In the case of Eqs. (3.4) or (3.2) the situation is more complicated still. The solution to the left of the initial data is sometimes referred to as the *backward continuation* of a solution. The problem of backward continuation is *incorrect in the Hadamard sense*. Consider again Eq. (3.5). Assume that the initial function $\varphi(t)$ is not known exactly; i.e., we know only that the function $\tilde{\varphi}(t) = \varphi(t) + \delta(t)$ with some estimate δ_0 of error $\delta(t)$, $\|\delta(t)\| \leq \delta_0$. Under these assumptions it is impossible to determine $x(t)$ to the left of $t = 0$ with precision δ_0 by the method described previously in this section. In fact, if we let $\delta(t) = \delta_0 \sin \omega t$, then $x(t) = \tilde{\varphi}(t) = \dot{\varphi}(t) + \delta_0 \omega \cos \omega t$ and $\|x(t) - \dot{\varphi}(t)\| = \delta_0 \omega \to \infty$ if $\omega \to \infty$. In this case it is necessary to apply a certain *regularization method* for backward continuation. Notice also

§3. Initial-Value Problem for Functional Differential Equations

that some applied problems require defining the initial state of the system from the known current state [107].

Let us give some known results [59, 81(4), 156(3)] about existence, uniqueness and continuous dependence of RFDE solutions with bounded delay:

$$\dot{x}(t) = f(t, x_t), \quad t \geq t_0, \quad x_{t_0} = \varphi \quad (3.6)$$
$$x_t = x(t + \theta), \quad -h \leq \theta \leq 0$$

A function $x(t)$ is said to be a solution of (3.6) on $[t_0, t_0 + \delta), \delta > 0$ if $x \in C[t_0 - h, t_0 + \delta)$ and $x(t)$ satisfies Eq. (3.6) for all $t \in [t_0, t_0 + \delta)$. We assume that $\dot{x}(t_0) = f(t, \varphi)$.

Theorem 3.1. *Suppose that $\Omega \subseteq R \times C[-h, 0]$ is open and that the map $f(t, \varphi)$ is continuous and satisfies a local Lipschitz condition in φ. If $(t_0, \varphi) \in \Omega$, then for some $\delta > 0$ there exists a unique solution of problem (3.6) on the interval $[t_0, t_0 + \delta)$. This solution depends continuously on initial data.*

3.3. RFDE with Unbounded Delay

Consider problem (3.2), (3.3) with unbounded delay. First assume that the initial function $\varphi(\theta) \in C[-\infty, 0]$. The definition of the solution is the same as in the case of bounded delay. Let the map $f(t, \varphi)$, $f: R_1 \times C[-\infty, 0] \to R_n$ be continuous and satisfy a local Lipschitz condition in

$$\varphi: |f(t, \varphi) - f(t, \psi)| \leq K\rho(\varphi, \psi)$$

for all φ, ψ from some subset $\Omega \subset C[-\infty, 0]$ and for all $t \in [t_0, t_0 + \delta)$. Under these assumptions, and using a fixed-point theorem, it is possible to prove local existence, uniqueness and continuous dependence (in the sense of metric ρ) of the solution of problem (3.2), (3.3) on initial data.

Other theorems can be obtained if the initial function is taken from a subspace of $C[-\infty, 0]$. For example, assume that $\varphi(\theta) \in CB[-\infty, 0]$ but that $CB[-\infty, 0]$ is a subspace of $C[-\infty, 0]$. Thus the set of continuous maps from $CB[-\infty, 0]$ into R_n is more powerful than the same set for $C[-\infty, 0]$. This fact implies that the conditions of existence and uniqueness for $\varphi \in CB[-\infty, 0]$ will be less restrictive than for $\varphi \in C[-\infty, 0]$. Other spaces of initial functions are considered in Corduneanu and Lakshmikantham [47]

3.4. NFDE

Investigations of NFDEs are of great interest but are more complicated than RFDEs. Consider the NFDE

$$\dot{x}(t) = f(t, x_t, \dot{x}_t), \quad t \geq t_0$$
$$\dot{x}(t) = \dot{x}(t + \theta), \quad x_t = x(t + \theta), \quad -\infty < \theta \leq 0 \quad (3.7)$$

An initial-value problem (Cauchy problem) for NFDE (3.7) can be stated as follows: Given an initial function φ and an initial moment t_0, find a solution $x: (-\infty, t_0 + \delta) \to R_n$ such that

$$x_{t_0} = x(t_0 + \theta) = \varphi(\theta), \qquad \dot{x}_{t_0} = \dot{x}(t_0 + \theta) = \dot{\varphi}(\theta) \quad -\infty < \theta \le 0 \quad (3.8)$$

The initial function $\varphi(\theta)$ in (3.8) is absolutely continuous. The simplest NFDE is a *neutral differential difference equation*

$$\dot{x}(t) = f[t, x(t), x(t-h), \dot{x}(t-h)] \qquad (3.9)$$

Here f is a given continuous function. The step method can be used in order to find the solution of (3.9). It is easy to verify that there does not exist a smooth solution even for the simplest NFDEs with an arbitrary smooth initial function. If the initial function satisfies a so-called *splicing condition* $\dot{\varphi}(0) = f[t_0, \varphi(\theta), \dot{\varphi}(\theta)]$, then the solution of (3.7), (3.8) has a continuous derivative. Solutions of NFDE (3.9) do not smooth, unlike solutions of RFDEs. Notice that if it is possible to obtain $\dot{x}(t-h)$ from (3.9), then the solution of (3.9), (3.8) always exists, not only to the right of initial interval $[t_0 - h, t_0]$, but also to the left of it.

The "splicing" condition is too restrictive for a number of problems. If it does not hold, then the solution of (3.7), (3.8) must be understood in some *generalized sense* [1, 253(2)]. Introduce a space $L_\infty[-h, 0]$ of functions $\varphi(\theta)$, $\varphi: R_1 \to R_n$, which are measurable and essentially bounded on $[-h, 0]$. Define the norm $\|\varphi\|_\infty$ in $L_\infty[-h, 0]$ by the formula $\|\varphi\|_\infty = \text{ess sup}|\varphi(\theta)|$, $-h \le \theta \le 0$. A function $x(t)$ is said to be a *solution of problem* (3.7), (3.8) if $x(t)$ is absolutely continuous on $[t_0 - h, t_0 + \delta)$ for $\delta > 0$, $\dot{x}(t) \in L_\infty[t_0 - h, t_0 + \delta)$ and $x(t)$ satisfies (3.7) and (3.8) almost everywhere. Another equivalent definition of solution of (3.7), (3.8) is given in Achmerov [1].

Assume that the map $f(t, \varphi, \psi)$, $f: R_1 \times C[-h, 0] \times L_\infty[-h, 0] \to R_n$ is continuous for some open set $\Omega \subseteq R_1 \times C[-h, 0] \times L_\infty[-h, 0]$ and is Lipshitzian in φ. Then

$$|f(t, \varphi_1, \psi) - f(t, \varphi_2, \psi)| \le L\|\varphi_1 - \varphi_2\|, \qquad (t, \varphi_1, \varphi_2) \in \Omega \quad (3.10)$$

The map $f(t, \varphi, \psi)$ satisfies a *modified Lipschitz condition* in ψ with the constant less than 1. It means that for any functions $\psi_1(\theta)$ and $\psi_2(\theta)$ such that $\psi_1(\theta) = \psi_2(\theta)$, $-h \le \theta \le -\varepsilon < 0$ for some $\varepsilon > 0$, we have

$$|f(t, \varphi, \psi_1) - f(t, \varphi, \psi_2)| \le K\|\psi_1 - \psi_2\|_\infty$$
$$K < 1, \qquad (t, \varphi, \psi_1), (t, \varphi, \psi_2) \in \Omega \quad (3.11)$$

Theorem 3.2 [253(2)]. *Let $\Omega \subseteq R_1 \times C[-h, 0] \times L_\infty[-h, 0]$ be open and let the map $f(t, \varphi, \psi)$ satisfy the hypothesis formulated above. If $(t_0, \varphi, \dot{\varphi}) \in \Omega$,*

§3. Initial-Value Problem for Functional Differential Equations

then there exists a unique solution of (3.7), (3.8) on the interval $[t_0, t_0 + \delta)$ for some $\delta > 0$. This solution depends continuously on initial data.

An interesting class of NFDE is

$$(d/dt)[x(t) - G(t, x_t)] = F(t, x_t), \quad t \geq t_0$$
$$x_{t_0}(\theta) = \varphi(\theta), \quad -h \leq \theta \leq 0 \tag{3.12}$$

Here maps $G, F: R_1 \times C[-h, 0] \to R_n$ are continuous. A function $x(t)$ is said to be a *solution of* (3.12) if $x(t) \in C[t_0 - h, t_0 + \delta)$, $\delta > 0$, $x(t)$ is equal to $\varphi(\theta)$ for $t \in [t_0 - h, t_0]$, the function $x(t) - G(t, x_t)$ is continuously differentiable and this function satisfies Eq. (3.12) for all $t \in [t_0, t_0 + \delta)$.

Theorem 3.3 [81(4), 83]. *Let $\varphi(\theta) \in C[-h, 0]$, and let the maps G and F be continuous in an open set $\Omega \subseteq R_1 \times C[-h, 0]$. The map $F(t, \varphi)$ is Lipschitzian in φ:*

$$|F(t, \varphi) - F(t, \psi)| \leq L\|\varphi - \psi\|, \quad (t, \varphi), (t, \psi) \in \Omega$$

The map $G(t, \varphi)$ satisfies in Ω a modified Lipschitz condition in φ with a constant less than 1:

$$|G(t, \varphi) - G(t, \psi)| \leq K\|\varphi - \psi\|, \quad K < 1$$

If $(t_0, \varphi) \in \Omega$, then there exists a unique solution of problem (3.12) for $t \in [t_0, t_0 + \delta)$, $\delta > 0$. This solution depends continuously on initial data.

Other existence and uniqueness theorems are also available [1, 5, 30, 81(4), 95(2)].

3.5. Two Useful Lemmas

Lemma 3.1. *Let a scalar function $x(t)$ satisfy an inequality $\dot{x}(t) \leq -\gamma x(t) + \beta\|x_t\|$, $t \geq t_0$, $x_t = x(t + \theta)$, $-h \leq \theta \leq 0$. Here γ, β are constants such that $\gamma > \beta \geq 0$. Then $|x(t)| \leq \|x_{t_0}\| \exp[-\lambda(t - t_0)]$, $t_0 \leq t$, where λ is a unique positive root of the equation $\lambda = \gamma - \beta \exp(\lambda h)$ [79(4)].*

Lemma 3.2 (Gronwall–Bellman inequality) [18]. *Let scalar continuous functions $x(t)$, and $g(t) \geq 0$ satisfy the inequality*

$$x(t) \leq \alpha(t) + \int_0^t g(s)x(s)\,ds, \quad t \geq 0$$

Here $\alpha(t)$ is a nondecreasing function. Then $x(t) \leq \alpha(t) \exp[\int_0^t g(s)\,ds]$.

For the proofs of Lemmas 3.1 and 3.2, see, for example, Kolmanovskii and Nosov [108(5)].

3.6. Estimations of Solutions Growth

Let us derive an *a priori* estimate on how the solution $x(t)$ of problem (3.7), (3.8) grows. Let the map f be measurable and let

$$|f(t, \varphi, \psi)| \leq C_1(1 + \|\varphi\|_\infty) + C_2\|\psi\|_\infty, \qquad 0 < C_2 < 1 \qquad (3.13)$$

with the initial function $\varphi(\theta) \in CB[-\infty, 0]$ and $\dot{\varphi}(\theta) \in L_\infty[-\infty, 0]$, $\|\varphi\|_\infty = \text{ess sup}|\varphi(\theta)|$, $-\infty < \theta \leq 0$.

Lemma 3.3. *If there exists a solution $x(t)$ of problem (3.7), (3.8), then for some positive constants C_3, C_4 the inequality holds:*

$$|x(t)| \leq C_3(1 + \|\varphi\|_\infty + \|\dot{\varphi}\|_\infty) \exp[C_4(t - t_0)], \qquad t \geq t_0 \qquad (3.14)$$

Proof. From (3.7), (3.8) it follows that

$$|\dot{x}(t)| \leq r_1(t) + C_2 r_2(t), \qquad t \geq t_0 \qquad (3.15)$$

$$r_1(t) = C_1(1 + \|\varphi\|_\infty) + C_2\|\dot{\varphi}\|_\infty + C_1 \sup|x(\tau)|, \qquad t_0 \leq \tau \leq t$$

$$r_2(t) = \sup|\dot{x}(t)|, \qquad t_0 \leq \tau \leq t$$

Since the right-hand part of the inequality (3.15) does not decrease, we have

$$r_2(t) \leq r_1(t) + C_2 r_2(t) \quad \text{or} \quad r_2(t) \leq (1 - C_2)^{-1} r_1(t) \qquad (3.16)$$

Integrating (3.7) from t_0 to t and taking into account (3.13), (3.15), and (3.16), we obtain

$$|x(t)| \leq |x(t_0)| + \int_{t_0}^t |f(s, x_s, \dot{x}_s)|\, ds$$

$$\leq |x(t_0)| + \int_{t_0}^t r_1(s)[1 + (1 - C_2)^{-1}]\, ds$$

The inequality (3.15) yields

$$\sup_{t_0 \leq \tau \leq t} |x(\tau)| \leq |x(t_0)| + (t - t_0)[1 + (1 - C_2)^{-1}]$$

$$\cdot \left[C_1(1 + \|\varphi\|_\infty) + C_2\|\dot{\varphi}\|_\infty + \int_{t_0}^t C_1[1 + (1 - C_2)^{-1}] \right.$$

$$\left. \cdot \sup_{t_0 \leq \tau \leq s}|x(\tau)|\, ds \right.$$

This inequality and Lemma 3.2 imply the estimate (3.14). Lemma 3.3 is proved.

3.7. Methods for Solving an FDE

This section is devoted to the brief discussion of certain methods of solution finding for an FDE. As mentioned previously, the step method is an effective method for solving some classes of FDEs. But its actual application is sometimes embarrassing, especially for large intervals. On the other hand, the ordinary differential equations arising in the step method are analytically integrable only in rare cases [57(2), 59, 156(3)].

Series development is another method of finding solutions [59, 152]. Some asymptotic methods may be used for an FDE with small delay [187(1), 187(2), 234, 235]. Applications of all of these methods are rather difficult and these methods are not universal.

The most universal method is *numerical integration of an initial-value problem*. The solution of problem (3.2), (3.3) is calculated in the following way. First we divide the interval $[t_0, T]$ into N parts, by the points $t_i = t_0 + i\tau$, $\tau = (T - t_0)N^{-1}$. The values x_i of the solution $x(t)$ at points $t_i = x(t_i)$ are defined by the relation

$$x_{i+1} = x_i + \tau f[t_i, x_i, x(t_i - h)] \qquad (3.17)$$

If the step-size τ for the numerical integration is chosen such that $m\tau = h$, where m is an integer, then relation (3.17) takes the form

$$x_{i+1} = x_i + \tau f(t_i, x_i, x_{i-m}); \qquad x_k = \varphi(-k\tau), \qquad k \leq 0 \qquad (3.18)$$

Recurrent relations (3.18) permit us to define the solution of problem (3.4), (3.3) on any time interval. For a survey of numerical methods for FDEs, see Cryer [49].

§4. LINEAR EQUATIONS

4.1. General Form of a Linear RFDE

Consider the linear RFDE

$$\dot{x}(t) = L(t, x_t) + f(t), \qquad x(t) \in R_n \qquad (4.1)$$

Here the operator $L: [0, \infty) \times C[-\infty, 0] \to R_n$ is continuous and linear in x_t; i.e., for any numbers α, β and functions φ, $\psi \in C[-\infty, 0]$ we will have $L(t, \alpha\varphi + \beta\psi) = \alpha L(t, \varphi) + \beta L(t, \psi)$. Using the Riesz theorem, Eq. (4.1) may be written in the following equivalent form [13, 186]

$$\dot{x}(t) = \int_0^\infty [d_s R(t, s)] x(t - s) + f(t) \qquad (4.2)$$

Here $R(t, s)$ is an $(n \times n)$ matrix, $R(t, 0) = 0$ and for any $t \geq 0$ there exists $\sigma(t)$ such that $R(t, s) = \text{const}$ for $\sigma(t) \geq s$. For all $t \geq 0$ the function $R(t, s)$ has a bounded variation $v(t)$ in s; the function $R[t, \sigma(t)]$ is continuous and $R(t, s)$ is mean continuous in t. If, in addition, $\sigma(t) \equiv h$, then we have an RFDE with bounded delay

$$\dot{x}(t) = \int_0^h [d_s R(t, s)] x(t - s) + f(t) \tag{4.3}$$

4.2. A Priori Solution Estimate

Obtain the estimate of the solution of Eq. (4.2) with the initial condition

$$x_{t_0} = \varphi(\theta), \qquad \varphi(\theta) \in CB[-\infty, 0] \tag{4.4}$$

Show that

$$|x(t)| \leq m(t) \leq \left[\|\varphi\|_B + \int_{t_0}^t |f(s)|\, ds \right] z(t)$$

$$z(t) = \exp\left[\int_{t_0}^t v(s)\, ds \right] \tag{4.5}$$

$$m(t) = \sup_{-\infty < s \leq t} |x(s, t_0, \varphi)|, \qquad t_0 \leq t < \infty$$

If $v(t)$ is a bounded function, then (4.5) coincides with (3.14). If $|x(s)| \leq \|\varphi\|_B$, $t_0 \leq s \leq t$, inequality (4.5) is obvious. Otherwise, denote $t_1 \in (t_0, t]$ such a point that $m(t) = |x(t_1)| > \|\varphi\|_B$. Integrating Eq. (4.2) we obtain

$$x(t) = x(t_0) + \int_{t_0}^t d\xi \int_0^\infty [d_s R(\xi, s)] x(\xi - s) + \int_{t_0}^t f(\xi)\, d\xi$$

Hence

$$m(t) = |x(t_1)| \leq \|\varphi\|_B + \int_{t_0}^t v(s) m(s)\, ds + \int_{t_0}^t |f(s)|\, ds \tag{4.6}$$

From (4.6) and Lemma 3.2, (4.5) follows. Inequality (4.5) means that the solution of (4.2), (4.4) is extendable on the semi-axis $[t_0, \infty)$.

Let us indicate some properties of RFDEs that distinguish them from ordinary differential equations. Usually the solution space of a linear RFDE is *infinite dimensional*, while for ordinary equations it is always finite dimensional. However, there are cases in which the solution space of an RFDE is finite dimensional for sufficiently large t; i.e., a part of the solution space is bound together [59, 156(4), 156(5)]. Further, in contrast to equations without delays, the solution of scalar linear homogeneous RFDEs with bounded coefficients *may decrease faster than does any exponent*. For example, the equation $\dot{x}(t) = -2tx(t-1)\exp(1-2t)$ has the solution $x(t) = \exp(-t^2)$, $t \geq 0$. The solutions of ordinary differential equations cannot decrease as quickly.

4.3. Linear NFDEs

Linear NFDEs have the form

$$\dot{x}(t) = L(t, x_t, \dot{x}_t) + f(t) \tag{4.7}$$

Here the continuous operator $L: [0, \infty) \times C[-h, 0] \times L_\infty[-h, 0] \to R_n$ is linear in x_t and $\dot{x}_t$. Like (4.3), Eq. (4.7) may be written as

$$\dot{x}(t) = \int_{-h}^{0} [d_s K_0(t, s)] x(t+s)$$
$$+ \int_{-h}^{0} [d_s K_1(t, s)] \dot{x}(t+s) + f(t) \tag{4.8}$$

Matrices K_0 and K_1 are assumed such that the corresponding operator $L(t, x_t, \dot{x}_t)$ is continuous. The initial-value problem (4.8), (3.8) may be stated for the NFDE as in §3. Theorem 3.2 implies the existence and uniqueness of the solution of problem (4.8), (3.8) under the assumptions

$$\sup_{t \geq t_0} \int_{-h}^{0} |d_s K_1(t, s)| < 1$$
$$\sup_{t \geq t_0} \int_{-h}^{0} |d_s K_0(t, s)| < \infty \tag{4.9}$$

Conditions (4.9) imply an *exponential growth estimate and continuity of solutions on semi-axis* $[t_0, \infty)$.

4.4. FDEs with Small Delays

An important class of FDEs are those equations with small delays. Consider Eq. (4.3) and assume that

$$v_0 = \sup_{t \geq t_0} v(t) = \sup_{t \geq t_0} \int_{0}^{h} |d_s R(t, s)| < \infty, \quad ev_0 h < 1 \tag{4.10}$$

Under conditions (4.10) for any t_0, x_0, there exists the unique two-sided solution $\bar{x}(t)$ of Eq. (4.3) defined on $-\infty < t < \infty$ and such that $|\bar{x}(t)| \leq C \exp(|t|/h)$, $\bar{x}(t_0) = x_0$ [57(2), 156(3), 187(1), 187(2)]. Such two-sided solutions are called *special solutions of Eq.* (4.3). Let $v_0 h < æ$, where $æ \in (0, e^{-1})$ is a root of the equation $1 + æ + \lambda(æ) = -\ln æ$ and $\lambda(æ)$ is the smallest root of the equation $\lambda = æ \exp(\lambda)$. Then for any solution $x(t)$ of Eq. (4.3), there exists the unique special solution $\bar{x}(t)$ of Eq. (4.3) such that

$$|x(t) - \bar{x}(t)| \leq C_1 \exp(-C_2 t); \qquad C_1 > 0, \quad C_2 > 0$$

Finally, special solutions of Eq. (4.3) are simultaneously the solutions of some ordinary differential equation. Thus, Eq. (4.3) with a small delay is *asymptotically equivalent* to an ordinary differential equation for $t \to \infty$. A similar result is valid for an NFDE with a small delay and a Lipshitz constant in $\dot{x}(t - s)$ smaller than one [59].

4.5. Variation-of-Constant Formula*

Obtain a formula for the solution of a linear FDE depending on equation coefficients and initial data. Consider first the equation

$$\dot{x}(t) = A(t)x(t) + B(t)x(t - h) + f(t), \qquad t \geq t_0, \quad x \in R_n$$

$$x(t_0 + \theta) = \varphi(\theta), \qquad -h \leq \theta \leq 0 \tag{4.11}$$

Let $y(t, s)$ be an $(n \times n)$ matrix

$$\frac{\partial y(t, s)}{\partial t} = A(t)y(t, s) + B(t)y(t - h, s), \qquad t \geq s$$

$$\frac{\partial y(t, s)}{\partial s} + y(t, s)A(s) + y(t, s + h)B(s + h) = 0, \qquad s \leq t \tag{4.12}$$

$$y(t, t) = I, \qquad y(t, s) = 0, \quad s > t$$

Multiplying Eq. (4.11) on $y(t, s)$ and integrating by parts, we get

$$x(t) = \int_{t_0}^{t} y(t, s)f(s)\, ds + \int_{t_0 - h}^{t} y(t, s + h)(s)\, ds + y(t, t_0)\varphi(t_0)$$

Obtain a similar formula for the general equation (4.8) under assumptions (4.9) and assume that $K_i(t, s) = K_i(t - h)$, $s \leq -h$, and $K_i(t, s) \equiv 0$, $s \geq 0$, $i = 0, 1$ [105(9), 108(5)]. Introduce the following $(n \times n)$ matrices

$$Q_1(\alpha, s) = K_0(s, \alpha - s)$$

$$Q_i(\alpha, s) = \int_{t_0}^{s} [d_t K_1(s, t - s)] Q_{i-1}(\alpha, t)$$

$$P_1(\alpha, s) = K_1(s, \alpha - s)$$

---
* As in the literature [14(1), 18, 79(3), 81(4), 108(5)].

§4. Linear Equations

$$P_i(\alpha, s) = \int_{t_0^-}^{s} [d_t K_1(s, t-s)] P_{i-1}(\alpha, t), \qquad i \geq 2$$

$$Q(\alpha, s) = \sum_{i=1}^{\infty} Q_i(\alpha, s)$$

$$P(\alpha, s) = \sum_{i=1}^{\infty} P_i(\alpha, s)$$

Here the notation t_0^- means that the point t_0 is excluded from the integration interval. Instead of (4.1.2), we introduce an $(n \times n)$ matrix $y(t, s)$, $y(t, t) = I$. The matrix $y(t, s)$ satisfies Eq. (4.8) with $f(t) = 0$ as a function of t for any fixed s and as a function of s for any fixed t

$$y(t, s) + \int_s^t y(t, \tau) Q(s, \tau)\, d\tau = I, \qquad s \leq t$$

$$y(t, s) = 0, \qquad s > t$$

Then for $t \geq t_0$ the following *representation formula* holds:

$$x(t) = y(t, t_0)\varphi(t_0) + \int_{t_0-h}^{t_0^-} \left[d_\alpha \int_{t_0}^{t} y(t, s) Q(\alpha, s)\, ds \right] \varphi(\alpha)$$

$$+ \int_{t_0-h}^{t_0} \left[d_\alpha \int_{t_0}^{t} y(t, s) P(\alpha, s)\, ds \right] \psi(\alpha) + \int_{t_0}^{t} y(t, s) f(s)\, ds$$

$$+ \int_{t_0^-}^{t} \left[d_\alpha \int_{t_0}^{t} y(t, s) P(\alpha, s)\, ds \right] f(\alpha) \tag{4.13}$$

Let Equation (4.8) be of the form

$$\dot{x}(t) = \sum_{i=1}^{m} B_i x(t - h_i) + B_0 \dot{x}(t - h) + f(t), \qquad t > 0 \tag{4.14}$$

The solution of problem (4.14), (4.4) will be

$$x(t) = y(t_0, t)\varphi(t_0) + \sum_{i=1}^{m} \int_{-h_i}^{0} \sum_{j=0}^{\beta-1} y(t, s + h_i + jh) B_i \varphi(s)\, ds$$

$$+ \int_{-h}^{0} \sum_{j=1}^{\beta} y(t, s + jh) B_0^j \psi(s)\, ds$$

$$+ \sum_{j=0}^{\beta-1} \int_{0}^{t} y(t, s + jh) B_0^j f(s)\, ds \tag{4.15}$$

where $\beta = [th^{-1}] + \chi(h - h[t/h])$ and $[t]$ is the greatest integer in t. The function $\chi(t) = 0$ for $t = 0$ and $\chi(t) = 1$ for $t > 0$. The matrix $y(t, s)$ in (4.15) as a function of t satisfies Eq. (4.14) with $f(t) = 0$, and as a function of s

$$\frac{\partial y(t, s)}{\partial s} + \sum_{i=1}^{m} \sum_{j=0}^{\beta-1} y(t, s + h_i + jh) B_0^j B_i = 0, \qquad s \leq t$$

$$y(t, t) = I; \qquad y(t, s) = 0, \qquad s > t \tag{4.16}$$

From (4.15), (4.16) it follows for the RFDE (4.14) with $B_0 = 0$ that

$$x(t) = y(t, t_0)\varphi(t_0) + \sum_{i=1}^{m} \int_{-h_i}^{0} y(t, s + h_i) B_i \varphi(s) \, ds$$

$$+ \int_{t_0}^{t} y(t, s) f(s) \, ds$$

where the $(n \times n)$ matrix $y(t, s)$ satisfies the equations

$$\frac{\partial y(t, s)}{\partial t} = \sum_{i=1}^{m} B_i y(t - h_i, s), \qquad s \leq t, \quad y(t, t) = I$$

$$\frac{\partial y(t, s)}{\partial s} + \sum_{i=1}^{m} y(t, s + h_i) B_i = 0, \qquad y(t, s) = 0, \quad s > t$$

At last for the RFDE (4.8), with $K_1 \equiv 0$ by virtue of (4.13), we have

$$x(t) = y(t, t_0)\varphi(t_0) + \int_{t_0}^{t} y(t, s) f(s) \, ds$$

$$+ \int_{t_0 - h}^{t_0} \left[d_\alpha \int_{t_0}^{t} y(t, s) K_0(s, \alpha - s) \, ds \right] \varphi(\alpha)$$

where the matrix $y(t, s)$ is

$$y(t, \alpha) + \int_{\alpha}^{t} y(t, s) K_0(s, \alpha - s) \, ds = I, \qquad \alpha \leq t$$

$$y(t, s) = 0, \qquad s > t$$

$$\frac{\partial y(t, s)}{\partial t} = \int_{-h}^{0} [d_s K_0(t, s)] y(t, s)$$

§5. LINEAR AUTONOMOUS EQUATIONS

5.1. Characteristic Function

One of the important and well-studied classes of FDEs is linear autonomous equations such as

$$\dot{x}(t) = \int_0^\infty [dK(s)]x(t-s) + f(t), \qquad x(t) \in R_n \qquad (5.1)$$

Here all elements of the $(n \times n)$ matrix $K(s)$ have bounded variation on $[0, \infty)$.

Consider a homogeneous linear autonomous RFDE

$$\dot{x}(t) = \int_0^\infty [dK(s)]x(t-s) \qquad (5.2)$$

We shall look for nontrivial solutions $x(t) = \alpha \exp(zt)$, where z is a complex number and $\alpha \in R_n$ a constant vector. Equation (5.2) has a nontrivial solution of the indicated form if and only if

$$J(z)\alpha = z\alpha, \qquad J(z) = \int_0^\infty e^{-zs}\, dK(s)$$

Thus the number z must be a solution of the characteristic equation

$$\det[zI - J(z)] = 0 \qquad (5.3)$$

Here I denotes the identity matrix. The function $D(z) = \det[zI - J(z)]$ is called a *characteristic function*.

5.2. Zeros of Characteristic Function

Indicate some properties of the function $D(z)$. Since for $\operatorname{Re} z \geq 0$,

$$\max_{1 \leq i \leq n} \sum_{j=1}^n \int_0^\infty |e^{-zs}\, dK_{ij}(s)| = \gamma < \infty$$

and then Eq. (5.3) has no zeros in the half-plane $\operatorname{Re} z > \gamma$. Further, if for any z,

$$\max_{1 \leq i \leq n} \sum_{j=1}^n \int_0^\infty |e^{-zs}\, dK_{ij}(s)| < \infty$$

then $D(z)$ is an entire function of z. If the matrix $K(s)$ is constant for $s \geq h \geq 0$, then from the theory of zeros distribution of entire functions [156(3)] it follows that either $D(z)$ is a polynomial or $D(z)$ has infinitely many zeros $z_1, z_2, \ldots, z_m, \ldots$, and therefore $\operatorname{Re} z_m \to -\infty$ for $m \to \infty$.

A number of papers are devoted to the study of several particular cases of Eq. (5.3) [18, 170, 253(4)]. The quasi-polynomials

$$D(z) = \sum_{1=0}^{m} \sum_{j=1}^{r(1)} a_{1j} z^1 \exp(-zj) \tag{5.4}$$

are studied more than the others. We call $a_{pq} z^p \exp(-zq)$ the *principal term of quasi-polynomial* (5.4) if $a_{pq} \neq 0$ and if for each other term $a_{mn} z^m \exp(-zn)$ we have $p > m$, $q > n$ or $p = m$, $q > n$ or $p > m$, $q = n$. Clearly, not every quasi-polynomial has a principal term. Pontryagin [175] has proved that the quasi-polynomial without a principal term always has infinitely many zeros with an arbitrarily large positive real part.

It is possible to describe more exactly the *distribution of zeros*. As a rule, the zeros are divided into those that are asymptotic and those that are nonasymptotic. Nonasymptotic zeros are situated near the origin. They can be found only numerically. For *asymptotic zeros* some general formulae can be found. First, consider some examples. It is easy to verify that the zeros z_k of $D(z) = a - \exp(z)$ equal

$$z_k = \ln|a| + i \arg a + 2k\pi i, \qquad i^2 = -1 \tag{5.5}$$

Now take the quasi-polynomial

$$D(z) = z - a \exp(-z) = 0, \qquad a = |a| \exp(i\alpha) \tag{5.6}$$

The graphs of functions z and $a \exp(-z)$ cannot have more than two real points of intersection. Consequently, all asymptotic zeros of (5.6) are complex. From (5.6), taking the logarithm we obtain

$$\operatorname{Re} z_k = \ln|a| - \ln|z_k|, \qquad k = 0, \pm 1, \pm 2, \ldots$$

$$\operatorname{Im} z_k = \alpha - \arg z_k + 2k\pi = 2k\pi + 0(1)$$

As $(\operatorname{Re} z_k)/|z_k| \to 0$, $k \to \pm \infty$, we have

$$|\operatorname{Im} z_k| = [|z|^2 - (\operatorname{Re} z_k)^2]^{1/2} = |z_k| + 0(1)$$

Thus, for asymptotic zeros of (5.6) we receive

$$z_k = \ln|a| - \ln|2k\pi| + i2k\pi + 0(1) \tag{5.7}$$

Formula (5.7) can be made more precise as

$$z_k = \ln|a| - \ln|2k\pi| + i[(2k - \tfrac{1}{2} \operatorname{sgn} k) + \alpha] + 0(\ln k/k) \tag{5.8}$$

Expression (5.8) shows that the distances between asymptotic zeros are greater than some positive number.

The asymptotic roots of the general equation (5.3), corresponding to one RFDE, are situated on a finite number of chains of the form (5.8). Therefore,

Re $z_k \to -\infty$ for $k \to \pm\infty$. But any sector containing the imaginary axis also contains all zeros of (5.3), except a finite number of them. Notice that the distances between zeros of general equation (5.3) can become infinitely small, always being positive [18, 100, 253(4)]

5.3. Representation of a Solution as an Infinite Series

We now discuss the representation of the solution of Eq. (5.2) as an infinite series of elementary solutions. Zubov [252(1)] has established the theorem on the *asymptotic property of such a series*. Any solution $x(t)$ of Eq. (5.2) can be connected with a series

$$x(t) \sim \sum_{k=1}^{\infty} p_k(t) \exp(z_k t) \tag{5.9}$$

Here $p_k(t)$ is a polynomial the degree of which is 1 less than the multiplicity of the root z_k of Eq. (5.3). If for some numbers $\varepsilon > 0$ and β the strip $\beta - \varepsilon < \mathrm{Re}\, z < \beta + \varepsilon$ has no zeros of Eq. (5.3), then the asymptotic formula holds:

$$x(t) = \sum_{\mathrm{Re}\, z_k > \beta} p_k(t) \exp(z_k t) + 0\,[\exp(\beta t)]$$

For a scalar RFDE of order n [18, 253(4)] it is proved that series (5.9) converges and can be differentiated n times term by term in this case. For systems of RFDEs, series (5.9) *does not necessarily converge*.

From Zubov's theorem follows an important corollary. If $\mathrm{Re}\, z_k \leq -\beta < 0$ (this condition is equivalent to $\mathrm{Re}\, z_k < 0$), then all solutions of the homogeneous equation (5.2) tend to zero for $t \to \infty$. In other words, the condition $\mathrm{Re}\, z_k < 0$ is *necessary and sufficient for asymptotic stability of RFDE (5.2)*.

5.4. Linear Autonomous NFDEs

Linear autonomous NFDEs have the form

$$\dot{x}(t) = \int_0^{\infty} [dK_0(s)] x(t-s) + \int_0^{\infty} [dK_1(s)] \dot{x}(t-s), \qquad t \geq 0 \tag{5.10}$$

All elements of the $(n \times n)$ matrices $K_0(s)$ and $K_1(s)$ have bounded variation on $[0, \infty)$ and $x(t) \in R_n$.

Clearly, the *characteristic equation*, corresponding to NFDE (5.10), is

$$\det\left[Iz - J(z) - z \int_0^{\infty} \exp(-zs)\, dK_1(s) \right] = 0 \tag{5.11}$$

The left-hand side of Eq. (5.11) is called a *characteristic function of NFDE (5.10)*. State some properties of Eq. (5.10) solutions with piecewise constant kernels $K_0(s)$ and $K_1(s)$, i.e., properties of neutral differential difference equations. In this case the characteristic function (5.11) has the form

$$D(z) = \sum_{l=0}^{m} \sum_{j=1}^{r(l)} a_{lj} z^l \exp(-zh_{lj}) \qquad (5.12)$$

For commensurable delays h_{mj} in terms corresponding to the highest derivative of Eq. (5.10), *asymptotic zeros of* (5.12) are situated on some chains of type (5.5) or (5.8). In the case of incommensurable delays h_{mj}, in addition to these chains there exists a finite number of strips that are parallel to the imaginary axis and which contain the zeros of (5.12). The real parts of the zeros fill in the corresponding intervals of the real axis densely everywhere. The imaginary parts of the zeros are almost asymptotically periodic. The multiplicity of each zero of (5.12) is finite and does not surpass $(1 + m) \max_{0 \leq l \leq m} r(l) + m$. The distribution of the zeros of the general characteristic function (5.11) cannot be studied practically.

If the highest derivative of a scalar NFDE enters only with *commensurable delays*, then any solution of such an equation can be represented as a *convergent infinite series of the type* (5.9). *Incommensurable delays* in the highest derivative series (5.9) *may diverge*. Take, for example, the equations

$$\dot{x}(t) - \dot{x}(t-1) - \dot{x}(t-h) - \dot{x}(t-1-h) = 0$$
$$x(t) = \varphi(t), \qquad x(t) = \dot{\varphi}(t), \qquad 0 \leq t \leq 1 + h \qquad (5.13)$$
$$\varphi(t) = (\tfrac{1}{2})(t-1)^2, \qquad 0 \leq t \leq 1; \qquad \varphi(t) = 0, \qquad 1 \leq t \leq 1 + h$$

In (5.13) the number h is defined as follows. Take any number $\alpha > 0$ and define the continued fraction

$$h = \frac{1}{q_0} + \frac{1}{q_1} + \frac{1}{q_2} + \frac{1}{q_3} + \cdots$$

The numbers q_k are given by the expressions $q_0 = 1$, $q_1 = 1, \ldots,$

$$q_{k+1} = [Q_k^{-1} \exp(2\pi\alpha Q_k) + 1], \ldots.$$

Here Q_k is a denominator of the convergent of a continued fraction:

$$P_k/Q_k = \frac{1}{q_0} + \frac{1}{q_1} + \frac{1}{q_2} + \cdots + \frac{1}{q_k}$$

and $[\beta]$ is the greatest integer in β. In Leontjev [125] it is shown that series (5.9) for Eq. (5.13) is divergent. The representation of the solution of the general equation (5.10) as an infinite series is not studied practically. For NFDE

$$\sum_{r=0}^{n} \sum_{s=0}^{m} a_{rs} x^{(r)}(t - h_s) = 0, \qquad a_{n0} \neq 0 \qquad (5.14)$$

under the assumption that Re $z_k \leq -\varepsilon < 0$, it is proved that all solutions approach zero more rapidly than an exponent; i.e., the trivial solution of Eq. (5.14) is *exponentially stable*. If only Re $z_k < 0$, then the situation is more complicated, since the NFDE can have a chain of zeros which approach the imaginary axis. There exists a scalar NFDE with one delay, the trivial solution of which is *asymptotically, but not exponentially, stable*. All solutions in this case have an exact estimation $x(t) = 0(t^{-\delta})$, $\delta > 0$ [78(1)].

In the case of a few delays, if there is a chain of multiple zeros which approach the imaginary axis from the left, the trivial solution may be stable or unstable [72(1), 72(2), 216]. The *supercritical case* in which Re $z_k \leq 0$ and there are infinitely many zeros on the imaginary axis is studied in Gromova [72(2)]. In this case the trivial solution may be stable or unstable.

§6. FEEDBACK SYSTEMS WITH AFTEREFFECTS

6.1. Single-Loop Feedback Systems

Feedback systems for the plants with aftereffect can be divided into two types: single-loop and two-loop systems [68, 214(1), 242(2)]. A structural scheme of a single-loop system is shown in Fig. 1.17. Assume that transfer function $W_p(s)$ of a plant with delay is

$$W_p(s) = H_p(s) \exp(-s\tau), \qquad H_p(s) = A(s)/B(s) \qquad (6.1)$$

Here $A(s)$ and $B(s)$ are two polynomials in s. Consider the feedback system shown in Fig. 6.1 in the case in which the controller is a standard one of the P,

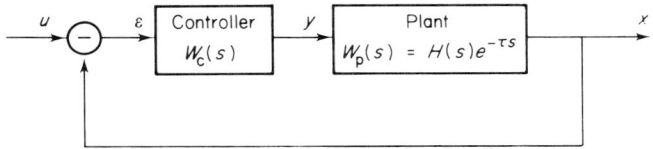

Fig. 1.17. Single-loop feedback system with delay.

PD, I, PI or PID type. Let the transfer function of the plant be

$$W_p(s) = K_p \exp(-s)/(Ts + q) \qquad (6.2)$$

Here K_p is a gain, T is a time constant and $\tau = 1$ is the delay of the plant. The number q can have two values: $q = 0$ or $q = 1$. Here $q = 0$ corresponds to an astatic plant with delay and $q = 1$ corresponds to an inertial plant. Air intake is an example of such a plant (see §1).

Table 1.2 lists some of the characteristics of a feedback system (6.2) closed by one of the standard controllers. In Table 1.2 we designate by $K = K_c K_p$ the gain of an open system by I the integral action time and D the derivative action time. For general plant (6.1) similar conclusions hold. If the degree of the polynomial $A(s)$ is one unit less than the degree of $B(s)$, then the type of FDE describing the closed system coincides with the type in Table 1.2. In the case in which the degree of $A(s)$ is two or more units less than the degree of $B(s)$, the closed system is governed by an RFDE. So, a closed feedback system for a plant with delay can be governed either by an RFDE or by an NFDE.

6.2. Effects of Little Delays

In the process of mathematical model construction one often neglects the existence of little delays. Sometimes this may lead to paradoxical conclusions [68, 75, 111, 133].

Consider one example of this type. Figure 1.18 shows an *ideal predictor* [108(5), 165] described by the equations $v(t) = K_1[u(t) + \varepsilon(t)]$, $\varepsilon(t) = K_2 v(t) + v(t - 1)$. These imply for $K_1 K_2 = 1$ that $u(t + 1) = -v(t)$. Thus the output at moment t depends on the input at a future moment $(t + 1)$. This contradicts the causality principle. The paradox can be explained by the

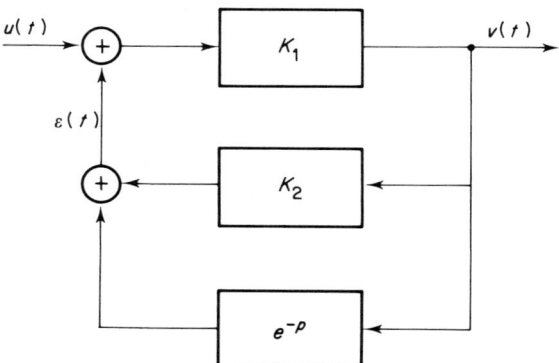

Fig. 1.18. "Ideal" predictor.

§6. Feedback Systems with Aftereffects

Table 1.2

No.	Controller type and its transfer function	Transfer function of a closed system	Equation of a closed system	Type of equation
1	Proportional (P): $W_c(s) = K_c$	$\dfrac{K_p \exp(-s)}{Ts + 1 + K \exp(-s)}$	$T\dot{x}(t) + x(t) + Kx(t-1)$ $= K_p y(t-1)$	Retarded
2	Proportional-plus-derivative (PD): $W_c(s) = K_c(1 + sD)$	$\dfrac{K_p \exp(-s)}{Ts + 1 + K(1+sD)\exp(-s)}$	$T\dot{x}(t) + KD\dot{x}(t-1) + x(t)$ $+ Kx(t-1) = K_p y(t-1)$	Neutral
3	Integral (I): $W_c(s) = K_c/(Is)$	$\dfrac{K_p Is \exp(-s)}{Is(Ts+1) + K \exp(-s)}$	$TI\ddot{x}(t) + I\dot{x}(t) + Kx(t-1)$ $= K_p I \dot{y}(t-1)$	Retarded
4	Proportional-plus-integral (PI): $W_c(s) = K_c(1 + 1/Is)$	$\dfrac{K_p Is \exp(-s)}{Is(Ts+1) + K - (Is+1)\exp(-s)}$	$TI\ddot{x}(t) - I\dot{x}(t)$ $+ KI\dot{x}(t-1) + Kx(t-1)$ $= K_p I \dot{y}(t-1)$	Retarded
5	Proportional-integral-derivative (PID): $W_c(s) = K_c(1 + Ds + 1/Is)$	$\dfrac{K_p Is \exp(-s)}{Is(Ts+1) + K(1 + Is + IDs^2)\exp(-s)}$	$TI\ddot{x}(t) + KID\ddot{x}(t-1)$ $+ KI\dot{x}(t-1) + I\dot{x}(t)$ $+ Kx(t-1) = K_p I \dot{y}(t-1)$	Neutral

effects of little delays. For example, if one allows a little delay h in unit K_2, then the predictor is described by the equations

$$v(t) = K_1[u(t) + \varepsilon(t)], \qquad \varepsilon(t) = K_2 v(t-h) + v(t-1)$$

This means that

$$v(t) = K_1[u(t) + u(t-h) + u(t-2h)]$$
$$+ K_1^2[u(t-1) + 2u(t-1-h)] + \cdots$$

Thus in this case the output is defined only by values of the input at preceding moments.

Another example of this type is absolutely invariant systems studied intensively by Schipanov et al. [196]. An absolutely invariant system loses its remarkable properties if one takes little delays into account.

6.3. Analysis of Single-Loop Feedback Systems

The knowledge of the transfer function permits us, with the aid of the inverse Laplace transform, to define the output of a feedback system. For example, the response $x(t)$ on the unit step input $1(t)$ for a closed system with $W_{cl}(s) = K_p \exp(-s)/[Ts + K \exp(-s)]$ is equal to

$$x(t) = \frac{K_p}{T} \sum_{j=1}^{[t]} \left(-\frac{K}{T}\right)^{j-1} \frac{(t-j)^j}{j!}, \qquad t > 1$$

Even in this simple case, the analytical expression is sufficiently complicated and is difficult to analyse. Graphical representation of transient processes is more convenient. Figure 1.19 represents transient processes in closed systems for different types of controllers and different time constants T. Notice that the steps of output in Fig. 1.19 may be explained by the presence of a δ function in corresponding NFDEs. The study of transient processes permits us to make some general conclusions. For all types of controllers the increase of delay τ (or the decrease of time constant T) leads to an increase of overshoot and of oscillating properties. The P controller may be used only for plants with a small delay, the properties of which are similar to those of plants without delay (see also §4). The use of PD controllers leads to a decrease of overshoot and of oscillating properties, but there is a position error. For I or PI controllers there is no position error, but the response time is sufficiently great. The best and most universal controllers are PID controllers. They result in a considerable decrease in overshoot and response time, but their adjustments are the most difficult [68, 188].

Let us now formulate some *general rules for choosing a controller*. P controllers give response times on the order of 5τ and a position error of less

§6. Feedback Systems with Aftereffects

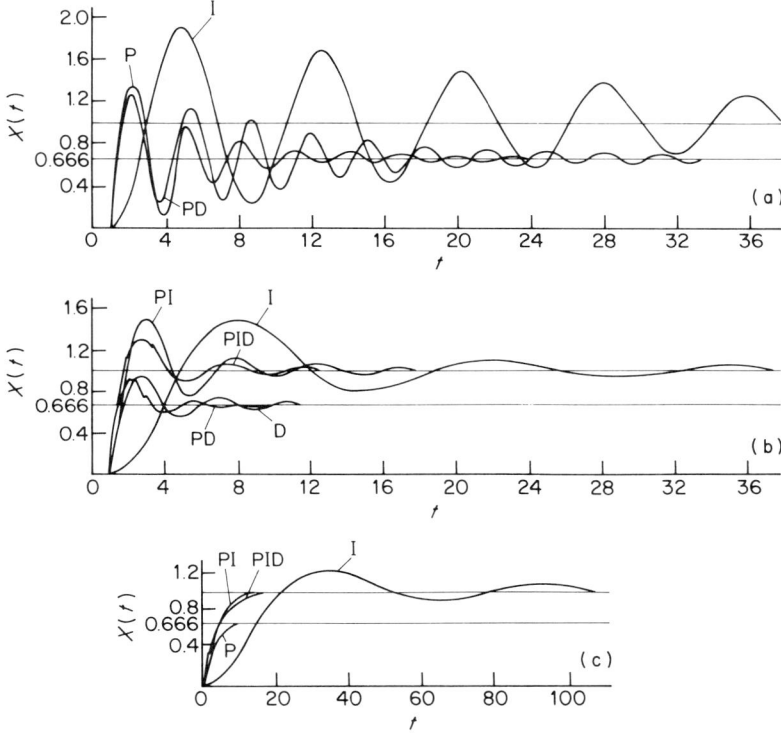

Fig. 1.19. Transient processes in systems with time lag. (Steps a–c.)

than 10% only for plants with minor delay $\tau < 0.05T$. PI controllers do not give position errors, but their response times are more than 6τ. The use of PID controllers permits us to decrease the response time to a time on the order of 4τ.

6.4. Two-Loop Systems with Delay (Smith, Reswick and Other Controllers)

Two-loop systems have good performances in comparison with single-loop ones. The *cascade circuit* shown in Fig. 1.20 can be used when the plant consists of two parts; the first one with transfer function $H_1(s)$ has no delay, and the second one has transfer function $H_2(s) \exp(-s\tau)$. The introduction of local feedback from the input of the second part permits us to compensate for disturbance f_1, but not for disturbance f_2. Thus the cascade circuit has a restrictive region of applications [12, 68].

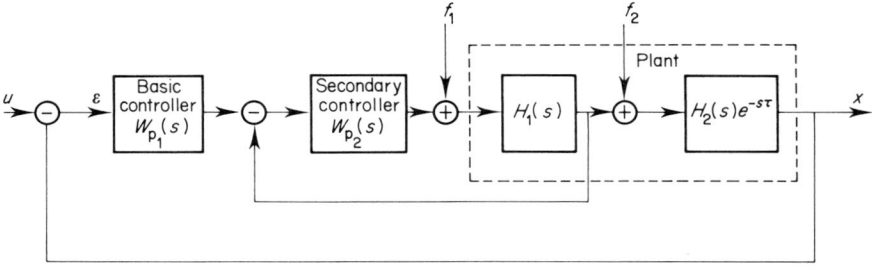

Fig. 1.20. Cascade circuit for a system with time lag.

The two-loop *system with delay compensation* shown in Fig. 1.21 is also considered [134(1), 134(2), 161]. Delay compensation can be used when the plant has not time lag, but the controller has a certain delay. Let us assume that a closed system containing only plant $H(s)$ and controller $W_c(s)$ has good performance. In the presence of unit $R(s) \exp(-s\tau)$, system performance grows worse, and some correction is necessary. For delay compensation, one introduces a second feedback containing an amplifier K_m and plant model $H_m(s)$ (see also Loginov *et al.* [131]).

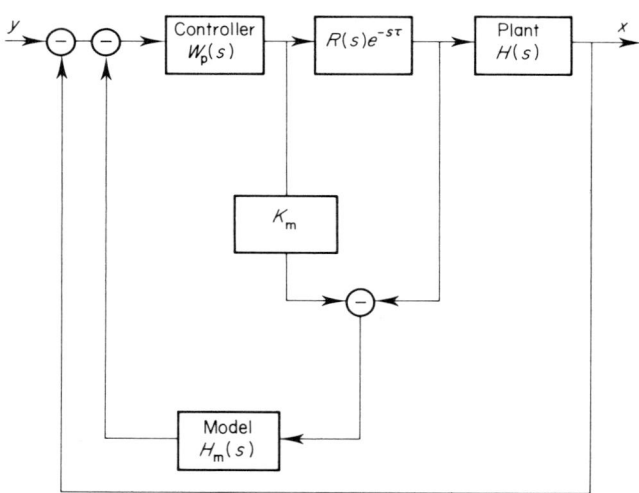

Fig. 1.21. System with delay compensation.

§6. Feedback Systems with Aftereffects

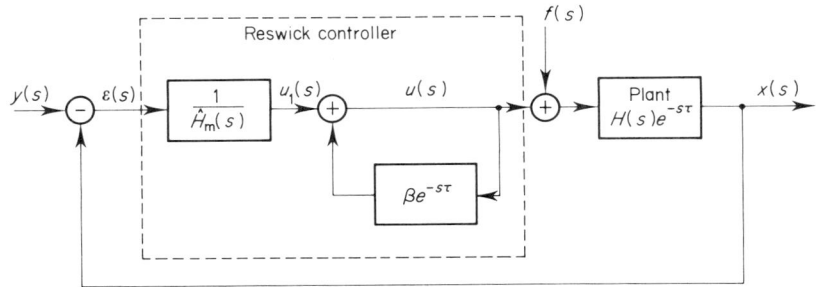

Fig. 1.22. Reswick controller for a system with delay.

A few computations show that the transfer function of a closed system equals

$$\Phi(s) = W_c(s)H(s)R(s)\exp(-s\tau)U^{-1}(s)$$
$$U(s) = 1 + K_m W_c(s)H(s) - W_c(s)R(s)\exp(-s\tau)H_m(s) + W_c(s)R(s)\exp(-s\tau)H(s)$$

For $H_m(s) = H(s)$ we have $U(s) = 1 + K_m W_c(s)H(s)$. Thus the unit $R(s)\exp(-s\tau)$ does not exert influence on the performance of the closed system.

In the case in which it is impossible to separate the delay from the rest of the plant, Smith or Reswick controllers are usually used. The basic idea of these controllers consists of the construction of a plant model which permits us to obtain plant output without delay. Consider the system containing the *Reswick controller*, shown in Fig. 1.22 [68, 184]:

$$\varepsilon(s) = y(s) - x(s), \qquad u_1(s) = \varepsilon(s)/H_m(s)$$
$$u(s) = u(s) + \beta u(s)\exp(-s\tau)$$
$$[u(s) + f(s)]H(s)\exp(-s\tau) = x(s)$$

Excluding from this system variables $\varepsilon(s)$, $u_1(s)$ and $u(s)$, we get

$$x(s) = H(s)\exp(-s\tau)\{y(s) + H_m(s)[1 - \beta \exp(-s\tau)]f(s)\}/U_1(s)$$
$$U_1(s) = H_m(s)[1 - \beta \exp(-s\tau)] + H(s)\exp(-s\tau)$$

For $H_m(s) = H(s)$ and $\beta = 1$ it follows that

$$x(s) = y(s)\exp(-s\tau) + H(s)[\exp(-s\tau) - \exp(-2s\tau)]f(s) \qquad (6.3)$$

Thus the delay is excluded from the denominator of the transfer function. The action of delay is reduced to the time shift of the input $y(t)$ and to the time shift 2τ of the transient processes caused by the disturbance $f(t)$. For

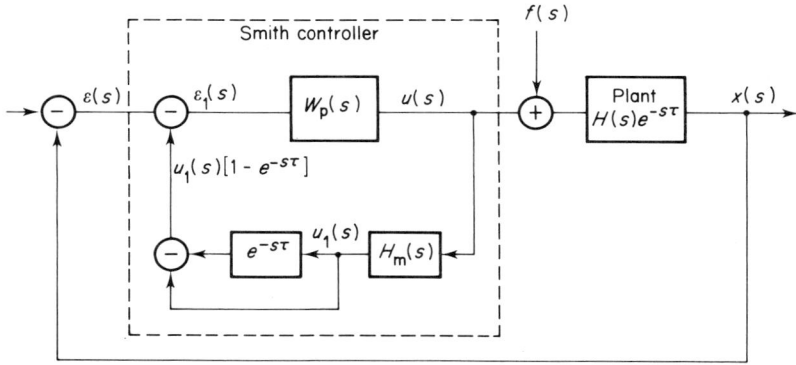

Fig. 1.23. Smith controller for a system with delay.

$H(s) = K_p$ the response time is equal to 2τ and is minimal. In fact, the disturbance can reach the controller input only by passing the plant, i.e., with delay τ. After a time τ the controller output signal reaches the plant output and can liquidate the error.

Practical realization of the Reswick controller is difficult because the simulation of unit $1/H(s)$ demands the exact knowledge of plant parameters. Also, it is necessary to take $\beta \cong 1$, but for $\beta > 1$ the closed system becomes unstable.

Let us now analyse the *Smith controller* [214(1), 214(2)]. A structural scheme (Fig. 1.23) yields

$$\varepsilon(s) = y(s) - x(s), \quad \varepsilon_1(s) = \varepsilon(s) - u_1(s)[1 - \exp(-s\tau)]$$
$$u(s) = W_p(s)\varepsilon_1(s), \quad u_1(s) = H_m(s)u(s) \quad (6.4)$$
$$x(s) = H(s)\exp(-s\tau)[u(s) + f(s)]$$

Excluding from (6.4) the auxiliary variables $\varepsilon(s)$, $\varepsilon_1(s)$, $u_1(s)$ and $u(s)$, we get

$$x(s) = [W_p(s)H(s)y(s)\exp(-s\tau)]/U_2(s)$$
$$+ \{H(s)\exp(-s\tau) + H(s)W_p(s)H_m(s)[\exp(-2s\tau)]f(s)\}/U_2(s)$$
$$U_2(s) = 1 + W_p(s)H(s)\exp(-s\tau) + W_p(s)H_m(s)[1 - \exp(-s\tau)]$$

Suppose that $H_m(s) = H(s)$. Then in the frequency strip where $|W_p(j\omega)H(j\omega)| \gg 1$, we obtain

$$x(s) \cong y(s)\exp(-s\tau) + H(s)[\exp(-s\tau) - \exp(-2s\tau)]f(s) \quad (6.5)$$

Relation (6.5) is just the same as (6.3). Thus, the Smith controller has the same performance as Reswick's. Systems containing Smith controllers have been studied elsewhere [68, 161, 190, 215]. The influence of parameter errors

§6. Feedback Systems with Aftereffects

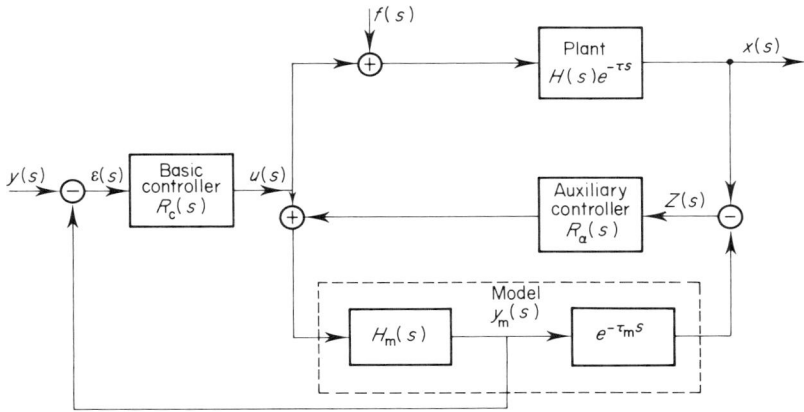

Fig. 1.24. Two-loop system with auxiliary controller.

between a plant and its model has also been investigated [94, 190, 215], and the results showed that the Smith controller provides good performances if the time constant error is less than 20% and the delay error is less than 10%. If disturbances act after the delay unit, then their action can be decreased by introducing into the controller a differentiator with transfer function $W_d(s) = Ts/(Ts + 1)$.

The main purpose of the *controller proposed by Kabal'nov et al.* [94] and shown in Fig. 1.24 is to limit the influence of parameter errors. Introduction of an auxiliary controller $R_a(s)$ permits us to conserve good performances for parameter errors on the order of 15%–20%.

If parameter errors are greater, then one can use the adaptive controller described in Chapter 2 §9.

Chapter 2

Stability of Retarded Equations

§1. GENERAL DEFINITIONS OF STABILITY

1.1. Definition of Liapunov Stability

Denote M a metric space of continuous functions $\varphi(\theta)$, $\varphi: (-\infty, 0] \to R_n$ with a metric ρ and Q_H a sphere in M (see Chapter 1, §3). Consider the initial-value problem

$$\dot{x}(t) = f(t, x_t), \qquad x_t(\theta) = x(t + \theta), \qquad -\infty < \theta \leq 0$$
$$x_{t_0}(\theta) = \varphi(\theta), \qquad \varphi(\theta) \in M, \qquad f: R_1 \times Q_H \to R_n \tag{1.1}$$

If the continuous map $f(t, \varphi)$ is Lipschitzian in φ, then there exists the unique solution $x(t, t_0, \varphi)$ of (1.1). In Liapunov stability theory we investigate the stability of a given solution. Let $z(t) \in Q_H$, $t_0 \leq t < \infty$ be a solution of problem (1.1).

Definition 1.1. *The solution $z(t)$ of problem (1.1) is called stable (in the Liapunov sense) if for any $\varepsilon \in (0, H)$ there exists $\delta = \delta(\varepsilon, t_0) > 0$ such that all solutions $x(t)$ of problem (1.1) satisfying the condition $\rho(x_{t_0}, z_{t_0}) \leq \delta$ are defined on $[t_0, \infty)$ and also $|x(t) - z(t)| \leq \varepsilon$ for $t \geq t_0$.*

Liapunov stability means that an arbitrarily narrow ε neighborhood of the solution $x(t)$ contains all those solutions of problem (1.1) which are sufficiently close to z_{t_0} at initial moment t_0. It should be noted that one must study the stability of a given solution $z(t)$ of problem (1.1) but not the stability of Eq. (1.1).

EXAMPLE 1.1 [36]. Consider the equation

$$\dot{x}(t) = -\tfrac{1}{2}[x^2(t) + (x^4(t) + 4x^2(t))^{1/2}]x(t) \tag{1.2}$$

The general solution of (1.2) is $x(t) = c \sin(xt + d)$, where c and d are constants. It is clear that the trivial solution $z(t) = 0$ of Eq. (1.2) is stable, but any other solution will be unstable. Actually, for the solutions $x_1(t) = c_1$

§1. General Definitions of Stability

$\sin(c_1 t + d_1)$ and $x_2(t) = c_2 \sin(c_2 t + d_2)$, where $c_1 \neq 0$, $c_2 \neq 0$ and c_2/c_1 is irrational, we have

$$\varlimsup_{t \to \infty} |x_1(t) - x_2(t)| = |c_1| + |c_2|$$

So the solutions $x_1(t)$ and $x_2(t)$ diverge on a finite value as $t \to \infty$ and this is independent from the proximity of $(x_1(t_0), \dot{x}_1(t_0))$ to $(x_2(t_0), \dot{x}_2(t_0))$.

REMARK 1.1. For linear equations $\dot{x}(t) = L(t, x_t) + g(t)$, $L(t, 0) = 0$ (where $L(t, \varphi)$ is a linear map) from the stability of a solution for some $g(t)$ follows the stability of any solution for arbitrary $g_1(t)$. Hence it is possible to call the linear equation stable or unstable.

The stability investigation of any solution $z(t)$ may be reduced to the study of stability of the trivial solution. For this it is sufficient to introduce the function $y(t) = x(t) - z(t)$. Then

$$\dot{y}(t) = f(t, y_t + z_t) - f(t, z_t) = g(t, y_t), \qquad g(t, 0) = 0 \qquad (1.3)$$

Below we assume that $f(t, 0) = 0$ and investigate only the stability of the trivial solution of (1.1).

Definition 1.2. *The trivial solution $x(t) = 0$ of Eq. (1.1) is called stable if for any $\varepsilon > 0$ there exists a $\delta = \delta(\varepsilon, t_0)$ such that $|x(t, t_0, \varphi)| \leq \varepsilon$, $t \geq t_0$ for any initial function $\varphi(\theta) \in Q_\delta$. Otherwise the trivial solution is called unstable.*

Let us make some remarks about the dependence of stability on the choice of initial moment t_0. Assume that sufficient conditions of existence, uniqueness and continuous dependence of solutions on initial data are fulfilled on $[s, t_0]$, $s \leq t_0$. Then, from the trivial solution stability related to disturbances of the initial function at the moment t_0, follows stability related to disturbances in any previous moment $t_1 \in [s, t_0]$. This statement is true by virtue of continuous dependence of the solution on $[s, t_0]$ from the initial data. At the same time the trivial solution stability related to disturbances at t_0 may be unstable related to disturbances at the moment $t_1 > t_0$. For ordinary differential equations there is no such effect because all solutions are left-continuable.

EXAMPLE 1.2 [253(1)]. Consider the equation

$$\dot{x}(t) = b(t)x(t - 3\pi/2) \qquad (1.4)$$

Here the continuous function $b(t) = 0$ for $t \leq 3\pi/2$; $b(t) = -\cos t$ for $3\pi/2 \leq t \leq 3\pi$; and $b(t) = 1$ for $t \geq 3\pi$.

Take $t_0 = 0$ and let φ be any continuous function: $\varphi \in C(-3\pi/2, 0]$. It is easy to see that the solution $x(t, 0, \varphi)$ equals $x(t, 0, \varphi) = \varphi(0)$ for $0 \leq t \leq 3\pi/2$ and $x(t, 0, \varphi) = -\varphi(0) \sin t$ for $t \geq 3\pi/2$. So $|x(t, 0, \varphi)| \leq |\varphi(0)|$ for $t \geq 0$.

Hence the trivial solution is stable related to disturbances at the moment $t_0 = 0$. Now consider another initial moment $t_1 \geq 3\pi$ and define the initial function $\varphi_1(t_1 + \theta) = \delta \exp[\lambda(t_1 + \theta)]$ for $-3\pi/2 \leq \theta \leq 0$. Here λ is a positive root of the equation $y = \exp(-3\pi y/2)$. We get that $x(t, t_1, \varphi_1) = \delta \exp[\lambda(t - t_1)]$ for $t \geq t_1 - 3\pi/2$ and also $|x(t, t_1, \varphi_1)| \to \infty$ as $t \to \infty$, $\delta \neq 0$. Thus the trivial solution is unstable related to the disturbances at any moment $t_1 \geq 3\pi$.

1.2. Uniform Stability

Definition 1.3. *The trivial solution of* (1.1) *is uniformly stable if for any $\varepsilon \in (0, H]$ there exists a $\delta(\varepsilon) > 0$ independent from t_0, φ such that $|x(t, t_0, \varphi)| \leq \varepsilon$ for $t \geq t_0$ and $\varphi(\theta) \in Q_\delta$, $t_0 \in R_1$.*

Uniform stability is stronger than the stability in the sense of Definition 1.2. Corresponding examples for nonautonomous equations are available [57(2), 108(5)]. However, for autonomous $(f(t, x_t) = f(x_t))$ or ω-periodic $(f(t + \omega, x_t) = f(t, x_t))$ Eq. (1.1) the uniform stability follows from the stability of the trivial solution. For any solution $x(t, t_0, \varphi)$ of ω-periodic Eq. (1.1) and any integer m we have

$$x(t, t_1, x_{t_1}(\theta, t_0, \varphi)) = x(t, t_0, \varphi), \qquad t_1 > t_0$$

$$x(t \pm m\omega, t_0 \pm m\omega, \varphi) = x(t, t_0, \varphi)$$

Hence for the proof of the uniform stability it is sufficient to investigate only the moments $t_0 \in [0, \omega]$. Take any $\varepsilon \in (0, H)$ and define $\eta(\varepsilon, \omega) > 0$ such that $|x(t, \omega, \varphi)| < \varepsilon$, $t \geq \omega$, $\|\varphi\| \leq \eta(\varepsilon, \omega)$. The existence of η follows from the stability for the initial moment $t_0 = \omega$. Choose a $\delta(\varepsilon)$ from Definition 1.3 such that

$$\max_{0 \leq t_0 \leq \omega} \max_{\|\varphi\| \leq \delta(\varepsilon)} |x(\omega, t_0, \varphi)| \leq \eta(\varepsilon, \omega)$$

By virtue of the continuous dependence of solutions at the moment $t = \omega$ on initial data it follows that there exists a $\delta(\varepsilon) > 0$ such that $|x(t, t_0, \varphi)| = |x(t, \omega, x_\omega(\theta, t_0, \varphi))| \leq \varepsilon$ for $t_0 \in [0, \omega]$ and $\|\varphi\| \leq \delta(\varepsilon)$.

1.3. Asymptotic Stability

Definition 1.4. *The trivial solution $x(t) = 0$ of* (1.1) *is called asymptotically stable if*: (1) *it is stable* (2) *for any t_0 there exists a $\Delta = \Delta(t_o) > 0$ such that*

$$\lim_{t \to \infty} x(t, t_0, \varphi) = 0, \qquad \varphi \in Q_\Delta \tag{1.5}$$

§1. General Definitions of Stability   47

The set $\Omega(t_0) \subseteq M$ of all initial functions φ such that $\lim x(t, t_0, \varphi) = 0, t \to \infty$, is called the attraction domain of the trivial solution at initial moment t_0.

Consider an example showing that in Definition 1.4 the first condition does not follow, in general, from the second one.

EXAMPLE 1.3. Given the equations

$$\dot{x}(t) = t^{-1}x(t) - 27t(t+2)^{-3}x^3((t+2)/3)$$
$$= t^{-1}x(t) - 27t(t+2)^{-3}x^3(t - \tau(t))$$
$$t \geq t_0 = 1, \quad \tau(t) = (2t-2)/3 \quad (1.6)$$

For $t \geq 1$ we have $\tau(t) \geq 0$. So (1.6) is the RFDE for $t \geq 1$ with initial data $x(1) = x_0$. The solution $x(t, 1, x_0) = x_0 t \exp(-x_0^2(t-1))$, $t \geq 1$, satisfies condition (1.5). But for $|x_0| = \delta$ and $t = 1 + \delta^{-2}$ we will have $|x(1 + \delta^{-2}, 1, x_0)| = \delta(1 + \delta^{-2}) \exp(-\delta^2(1 + \delta^{-2})) = (\delta + \delta^{-1})/e \geq 2/e$. Hence, the first condition of Definition 1.4 is not fulfilled. The trivial solution of (1.6) is not stable.

Definition 1.5. *The trivial solution of* (1.1) *is called uniformly asymptotically stable if it is uniformly stable and for any $\gamma > 0$ there exist $T(\gamma) > 0$ and $H_1 \in (0, H)$ such that $|x(t, t_0, \varphi)| \leq \gamma$ for $t \geq t_0 + T(\gamma)$ and $\varphi(\theta) \in Q_{H_1}$, $t_0 \in R_1$. In addition, the sphere $Q_{H_1} \subseteq M$ lies in the domain of attraction of the trivial solution.*

Uniform asymptotic stability is stronger than asymptotic stability. But if Eq. (1.1) is autonomous or ω periodic and its trivial solution is asymptotically stable, then it will be uniformly asymptotic stable.

1.4. Other Stability Definitions. Stability Under Delay Disturbances

Besides the previously mentioned stability definitions one sometimes uses the definitions of *global stability, stability under steady acting disturbances, exponential stability, partial stability, stability on finite time interval*, etc. For a detailed discussion of the different stability definitions see Kolmanovskii [108(5)], Malkin [138] and Yoshizawa [249]. However, let us remark that for a linear RFDE, exponential stability follows from uniform asymptotic stability. A specific problem for differential-difference equations is the *stability under delay disturbances* [129, 212, 223(2)]. For linear equations, some involving results are given in El'sgol'tz [58(2)] and Kolmanovskii and Nosov [108(5)]. If the trivial solution of a nonlinear RFDE is uniformly asymptotically stable for a given delay, then it is stable for any delays whose distances from the given delay are sufficiently small [114(3)–114(5)]. In general, this statement is not valid for NFDEs.

The following problem is interesting for equations with variable delay. Assume that a linear system is asymptotically stable for every constant delay from a given interval. Will this system be stable for variable delay from the same interval? Consider an example showing that it is not true in general.

EXAMPLE 1.4 [156(23)]. Investigate the scalar equation

$$\dot{x}(t) = -x(t - \tau(t)), \quad x(0) = x_0, \quad t \geq 0 \quad (1.7)$$

Take a number $a \in (\frac{3}{2}, \pi/2)$. Let the delay $\tau(t)$ in (1.7) be constant: $\tau(t) = \tau = $ const. Then for $0 \leq \tau \leq a$ the trivial solution of (1.7) is asymptotically stable (see Fig. 2.1). Define the variable delay $\bar{\tau}(t) \in [0, a]$

$$\bar{\tau}(t) = \begin{cases} t, & 0 \leq t \leq a \\ a, & a \leq t \leq a + 1 \end{cases}$$

For $t \geq a + 1$ we continue $\bar{\tau}(t)$ as a periodic function of the period $a + 1$, i.e., $\bar{\tau}(t + (a + 1)) = \bar{\tau}(t)$. By the method of steps we get for the solution of (1.7) with $\tau = \bar{\tau}(t)$ that $x(k(a + 1)) = x_0(\frac{1}{2} - a)^k$. But $|\frac{1}{2} - a| > 1$. Hence $|x(k(a + 1))| \to \infty$ for $k \to \infty$. So the trivial solution of (1.7) with $\bar{\tau}(t)$ is unstable. Notice that the trivial solution of the scalar equation $\dot{x}(t) = -a(t)x(t - \tau(t))$, where $a(t) > 0$ and $\tau(t) \geq 0$, is asymptotically stable if $(\sup_t a(t))(\sup_t \tau(t)) < \frac{3}{2}$ [156(23)].

§2. LINEAR AUTONOMOUS RETARDED EQUATIONS

2.1. Statement of the Problem

Stability of linear autonomous RFDEs has been studied by many authors [16, 18, 59, 252(1)] with the aid of a series expansion of solutions. But usually the arrangement of the characteristic quasi-polynomial roots is investigated in detail only for differential-difference equations. For some results about convergence of expansion in series of basic solutions see §1.5. Krasovskii [114(5)] considered an RFDE with finite delay using semi-group theory. In this paragraph, on the basis of Laplace transform, stability conditions for a linear autonomous RFDE with arbitrary delay are ascertained.

Consider the RFDE

$$\dot{x}(t) = \int_0^\infty [dK(s)]x(t - s), \quad x(t) \in R_n, \quad t \geq 0 \quad (2.1)$$

$$x(\theta) = \varphi(\theta), \quad -\infty < \theta \leq 0 \quad (2.2)$$

§2. Linear Autonomous Retarded Equations

Assume that all elements K_{ij} of the matrix $K(s)$ are functions with bounded variations on $[0, \infty)$ and the integral in (2.1) is a Riemann–Stiltjes one. Moreover

$$\int_0^\infty s |dK_{ij}(s)| < \infty, \qquad i, j = 1, \ldots, n \tag{2.3}$$

The initial function φ is continuous and is such that

$$\sup_{t \geq 0} |F(t)| < \infty, \qquad F(t) = \int_t^\infty [dK(s)] \varphi(t - s) \tag{2.4}$$

Introduce the norm $\|\varphi\|_0$ in the space of initial functions satisfying (2.4) and

$$\|\varphi\|_0 = |\varphi(0)| + \left(\int_0^\infty |F(t)|^2 \, dt \right)^{1/2}, \qquad \int_0^\infty |F(t)|^2 \, dt < \infty \tag{2.5}$$

Here $|\varphi(0)|$ denotes the euclidian norm in R_n.

Conditions (2.3), (2.4) are sufficient for existence, uniqueness and continuous dependence of solutions $x(t, \varphi)$ to problem (2.1), (2.2) on initial data. The following estimate also holds

$$|x(t, \varphi)| + |\dot{x}(t, \varphi)| \leq C_1 \|\varphi\|_0 \exp(C_2 t) \tag{2.6}$$

In (2.6) and hereafter, C_i denote some positive constants. Hence, there exists a Laplace transform for the solution $x(t, \varphi)$.

Definition 2.1. *The trivial solution $x(t) = 0$ of problem (2.1), (2.2) is called L_2 stable if for the solution $x(t, \varphi)$ we get*

$$\int_0^\infty |x(t, \varphi)|^2 \, dt < \infty$$

for any φ, $\|\varphi\|_0 < \infty$.

Stability and asymptotic stability definitions for Eq. (2.1) coincide with Definitions 1.3 and 1.4 with $\|\varphi\|_0$ instead of ρ. Find the Laplace transform of the problem (2.1), (2.2) solution. From (2.1) it follows that

$$\int_0^\infty e^{-zt} \dot{x}(t) \, dt = \int_0^\infty e^{-zt} \, dt \int_0^\infty [dK(s)] x(t - s) \tag{2.7}$$

Integrating by parts,

$$\int_0^\infty e^{-zt} \dot{x}(t) \, dt = z \bar{x}(z) - x(0)$$

$$\bar{x}(z) = \bar{x}(z, \varphi) = \int_0^\infty e^{-zt} x(t, \varphi) \, dt \tag{2.8}$$

Using the Fubini theorems,

$$\int_0^\infty e^{-zt} dt \int_0^\infty [dK(s)]x(t-s) = \int_0^\infty F(t)e^{-zt} dt$$

$$+ \int_0^\infty [dK(s)] \int_0^\infty e^{-zt} x(t-s) dt = \bar{F}(z) + \bar{K}(z)\bar{x}(z)$$

Here

$$\bar{F}(z) = \int_0^\infty F(t)e^{-zt} dt, \qquad \bar{K}(z) = \int_0^\infty [dK(s)]e^{-zs}$$

Substitute this expression in (2.7). Then

$$[Iz - \bar{K}(z)]\bar{x}(z) = x(0) + \bar{F}(z) \tag{2.9}$$

where I is the unit matrix.

The expression

$$\Delta(z) = \det[Iz - \bar{K}(z)] \tag{2.10}$$

is called a *characteristic function*, corresponding to Eq. (2.1).

2.2. General Theorems

Theorem 2.1. *Let the characteristic function* (2.10) *have no zeros in the half-plane* $\operatorname{Re} z \geq 0$ *and the kernel* $K(s)$ *satisfy condition* (2.3). *Then the trivial solution of Eq.* (2.1) *is* L_2 *stable.*

Proof. By virtue of inversion theorem [123] we have for some $\gamma > 0$

$$x(t, \varphi) = \frac{1}{2\pi i} \int_{\gamma - i\infty}^{\gamma + i\infty} \bar{x}(z, \varphi) e^{zt} dt$$

Under the assumptions of Theorem 2.1 the function $\bar{x}(z)$ is analytic for $\operatorname{Re} z \geq 0$. Actually,

$$\bar{x}(z) = [Iz - \bar{K}(z)]^{-1}(x(0) + \bar{F}(z))$$

Further, the matrix $[Iz - \bar{K}(z)]^{-1}$ is analytic for $\operatorname{Re} z > 0$ and continuous for $\operatorname{Re} z \geq 0$ since function (2.10) does not vanish and $\bar{K}(z)$ by virtue of (2.3) is bounded for $\operatorname{Re} z \geq 0$. Hence, using the Cauchy theorem [123],

$$x(t, \varphi) = \frac{1}{2\pi} \int_{-\infty}^{\infty} \bar{x}(i\beta) e^{i\beta t} d\beta \tag{2.11}$$

§2. Linear Autonomous Retarded Equations

By virtue of Plancherel theorem [186] it is sufficient for L_2 stability that

$$\int_{-\infty}^{\infty} |\bar{x}(i\beta)|^2 \, d\beta < \infty, \qquad \|\varphi\|_0 < \infty \tag{2.12}$$

Show (2.12). Since function (2.10) has no zeros on imaginary the axis and the elements of the matrix $\bar{K}(z)$ are bounded on it,

$$\|(i\beta I - \bar{K}(i\beta))^{-1}\| \leq 2C|\beta|^{-1}, \qquad |\beta| \to \infty$$
$$\|(i\beta I - \bar{K}(i\beta))^{-1}\| \leq C_1, \qquad -\infty < \beta < \infty \tag{2.13}$$

From (2.13)

$$\int_{-\infty}^{\infty} |(Ii\beta - \bar{K}(i\beta))^{-1} x(0)|^2 \, d\beta < \infty \tag{2.14}$$

From (2.13) and the Plancherel theorem

$$\int_{-\infty}^{\infty} |(i\beta I - \bar{K}(i\beta))^{-1} \bar{F}(i\beta)^2| \, d\beta \leq C_1^2 \int_{-\infty}^{\infty} |\bar{F}(i\beta)|^2 \, d\beta$$
$$= C_1^2 2\pi \int_0^{\infty} |F(t)|^2 \, dt \leq C_1^2 2\pi \|\varphi\|_0^2 \tag{2.15}$$

Using (2.9), (2.14) and (2.15) we obtain (2.12). Theorem 2.1 is proved.

Theorem 2.2. *Under the assumptions of Theorem 2.1 the trivial solution of (2.1) is asymptotically stable.*

Proof. Show that for $\|\varphi\|_0 < \infty$

$$x(t, \varphi) \to 0, \qquad t \to \infty \tag{2.16}$$

From (2.11) it follows that $x(t, \varphi)$ may be considered as a Fourier coefficient for $\bar{x}(i\beta)$. Hence, by virtue of one theorem on Fourier series ([61], p. 529), relation (2.16) holds if each component $\bar{x}_j(i\beta)$ may be presented as a sum of two functions. The first of them is absolutely integrable on $(-\infty, \infty)$ and the second one is bounded, monotonic for all sufficiently large $|\beta|$ and tends to zero for $|\beta| \to \infty$. On the basis of the Plancherel theorem, (2.6) and (2.13),

$$\int_{-\infty}^{\infty} |(i\beta I - \bar{K}(i\beta))^{-1} \bar{F}(i\beta)| \, d\beta$$
$$\leq \left(\int_{-\infty}^{\infty} |\bar{F}(i\beta)^2 \, d\beta \right)^{1/2} \left(\int_{-\infty}^{\infty} \|i\beta I - \bar{K}(i\beta)\|^{-2} \, d\beta \right)^{1/2}$$
$$\leq 2\pi \left(\int_0^{\infty} |F(t)|^2 \, dt \right)^{1/2} \left(\int_{-\infty}^{\infty} \|i\beta I - \bar{K}(i\beta)\|^{-2} \, d\beta \right)^{1/2} < \infty$$

$$\|K\| = \sum_{i,j=1}^{n} |K_{ij}|$$

This yields

$$\int_{-\infty}^{\infty} |(i\beta I - \bar{K}(i\beta))^{-1} \bar{F}(i\beta)| \, d\beta < \infty$$

Consider the function

$$q_j(\beta) = [(i\beta I - \bar{K}(i\beta))^{-1} x(0)]_j$$

$$= \sum_{l=0}^{n} \frac{A_{lj}(i\beta)}{\Delta(i\beta)} x_l(0)$$

$$= \sum_{l=0}^{n} \frac{a_{lj}(i\beta)^{n-1} x_l(0)}{\Delta(i\beta)} + O(|\beta|^{-2}) \qquad (2.17)$$

Here $A_{lj}(i\beta) = a_{lj}(i\beta)^{n-1} + O(|\beta|^{n-2})$, $(j, l = 1, \ldots, n)$ are cofactors of the matrix $(i\beta I - \bar{K}(i\beta))$. If $\sum_{l=0}^{n} a_{lj} x_l(0) = 0$, then because of (2.17),

$$|q_j(\beta)| \leq C_4 |\beta|^{-2}, \qquad \int_{-\infty}^{\infty} |q_j(\beta)|^2 \, d\beta < \infty$$

If $\sum_{l=0}^{n} a_{lj} x_l(0) \neq 0$, then from (2.17) it follows that $q_j(\beta)$ is bounded and for large $|\beta|$:

$$q_j(\beta) = (i\beta)^{n-1} \sum_{l=0}^{n} \frac{a_{lj} x_l(0)}{(i\beta)^n + O(|\beta|^{n-1}) + O(|\beta|^{-2})}$$

$$= \left[\sum_{l=0}^{n} \frac{a_{lj} x_l(0)}{(i\beta) + O(|\beta|^{-2})} \right] \qquad (2.18)$$

Hence $|q_j(\beta)|$ is expanded in the sum of two functions. One of them for large $|\beta|$ is equal to $|\beta|^{-1}|\sum_{l=0}^{n} a_{lj} x_l(0)|$ and another is absolutely integrable. So (2.16) follows from the work of Fichtengol'tz [61].

Show now that the trivial solution is stable. Define the operator sequence $\{A_n\}$ mapping the space of the initial function $\varphi(\theta)$ with norm (2.6) into the space of continuous and bounded functions: $A_n \varphi = x(t, \varphi)$ for $0 \leq t \leq n$ and $A_n \varphi = x(n, \varphi)$ for $t \geq n$. Every operator A_n is continuous on the basis of solution continuous dependence on initial data. From (2.16) we get that every solution $x(t, \varphi)$ of the problem (2.1), (2.2) with $\|\varphi\|_0 < \infty$ is bounded. Hence, the sequence $\{A_n\}$ is bounded on every element φ; i.e., $\sup_n A_n \varphi < \infty$. Using now the concentration of singularity principle ([96], p. 269), we obtain that the sequence $\{A_n\}$ is uniformly bounded in n:

$$\|A_n \varphi\|_B = \sup_{n \geq t \geq 0} |x(t, \varphi)| \leq \gamma \|\varphi\|_0 \qquad (2.19)$$

where the constant γ is independent of n. In the linear case, stability of the trivial solution follows from (2.19). Theorem 2.2 is proved.

Theorem 2.3. *Let for some $\gamma > 0$*

$$\int_0^\infty e^{\gamma s} |dK_{ij}(s)| < \infty, \qquad i, j = 1, \ldots, n$$

and also let the characteristic function (2.10) have no zeros in the half-plane Re $z \geq -\gamma$. Then $|x(t, \varphi)| \leq C_\varphi \exp(-\gamma t)$.

Proof. Consider the function $y(t) = x(t) \exp(\gamma t)$. Substituting $y(t)$ in (2.1) we obtain the characteristic function $\Delta_\gamma(z) = \det[(z - \gamma)I - \bar{K}(z - \gamma)]$ of the equation for $y(t)$. This and the assumptions of Thorem 2.3 imply that $\Delta \gamma(z)$ has no zeros in the half-plane Re $z \geq 0$. So by virtue of Theorem 2.2 we get $\lim_{t \to \infty} y(t) = 0$. Theorem 2.3 is proved.

§3. METHODS OF STABILITY INVESTIGATION OF LINEAR AUTONOMOUS RFDEs

By virtue of Theorem 2.2 the stability of RFDE (2.1) depends on the presence of zeros of the characteristic functions (2.11) in the half-plane Re $z \geq 0$. We shall call "p-zeros" the zeros with positive real parts. For ordinary differential equations, function (2.11) is reduced to a polynomial. Conditions for absence of p-zeros for polynomials are given by the well-known Routh–Hurwitz theorem. For general characteristic function (2.11) such complete results are not obtained.

3.1. Analytical Methods (Pontriagin's and Chebotarev's Theorems)

Consider the quasi-polynomial

$$D(z) = \sum_{l=0}^{m} \sum_{j=1}^{r} a_{lj} z \exp(b_j z) \tag{3.1}$$

Quasi-polynomials (3.1) are characteristic functions for differential-difference equations. The Routh–Hurwitz problem for quasi-polynomials (3.1) may be studied by different methods. Let us examine some of them. Quasi-polynomial with commensurable delays may be presented as

$$D_1(z) = \sum_{l=0}^{m} \sum_{j=1}^{r} a_{lj} z^l \exp(zj) \tag{3.2}$$

Consider the quasi-polynomial $D_1(z)$ for $z = i\omega$, where ω is a real number:
$D_1(i\omega) = g(\omega) + if(\omega)$.

Pontriagin's Theorem [175]. *If quasi-polynomial (3.2) has no p-zeros then all the zeros of the functions $g(\omega)$ and $f(\omega)$ are real, simple and alternating and*

$$\dot{g}(\omega)f(\omega) - \dot{f}(\omega)g(\omega) > 0, \qquad -\infty < \omega < \infty \qquad (3.3)$$

For absence of p-zeros of $D_1(z)$ one of the following conditions is sufficient:

(1) *All the zeros of the functions $g(\omega)$ and $f(\omega)$ are real, simple and alternating and inequality (3.3) is fulfilled for at least one real ω;*

(2) *all the zeros of the function $g(\omega)$ [or $f(\omega)$] are real and simple and for each zero relation (3.3) is satisfied.*

The simple proof of Pontriagin's theorem is given by Postnikov [177]. Consider, as example the equation,

$$\dot{x}(t) = bx(t-h), \qquad b > 0 \qquad (3.4)$$

The quasi-polynomial $D(z) = z\exp(hz) - b = 0$ corresponds to Eq. (3.4). By setting $z = i\omega$ we get

$$D(i\omega) = \omega\sin\omega h - b + i\omega\cos\omega h = g(\omega) + if(\omega)$$

$$g(\omega) = \omega\sin\omega h - b, \qquad f(\omega) = \omega\cos\omega h.$$

Zeros of the function $f(\omega) = \omega\cos\omega h$ are equal to $\omega_0 = 0$, $\omega_{k+1} = (k\pi + \pi/2)/h$; $k = 0, 1, \ldots$. These zeros are real and simple. Relation (3.3) for $\omega = \omega_k$ has the form $\dot{g}(0)f(0) - \dot{f}(0)g(0) = b > 0$, $\dot{g}(\omega_{k+1})f(\omega_{k+1}) - \dot{f}(\omega_{k+1})g(\omega_{k+1}) = \omega_{k+1}h(\omega_{k+1} - b) > 0$. The last inequality will be fulfilled under the conditions

$$b < \omega_1 < \cdots < \omega_{k+1}$$

So the trivial solution of Eq. (3.4) is asymptotically stable for $2bh < \pi$. By analogy we may prove that the equation $\dot{x}(t) = ax(t) + bx(t-h)$ is asymptotically stable if $a < h^{-1}$, $a < -b < (a^2 + p^2h^{-2})^{1/2}$, where p is the root of equation $p = ah^{-1}\mathrm{tg}p$ such that $0 < p < \pi$. If $a = 0$ we set $p = \pi/2$.

One of the ways of studying quasi-polynomials (3.1) with incommensurable delays is *generalization of the Routh–Hurwitz conditions*. Expand the quasi-polynomial $D(z)$ in the series $D(z) = a_0 + a_1z + a_2z^2 + \cdots$ and define the functions $u(z)$ and $v(z)$:

$$D(iz) = u(z) + iv(z), \qquad u(z) = a_0 - a_2z^2 + a_4z^4 + \cdots$$

$$v(z) = a_1z - a_3z^3 + a_5z^5 - \cdots$$

§3. Methods of Stability Investigation of Linear Autonomous RFDEs

Let us introduce the determinants Q_m:

$$Q_1 = a_1, \quad Q_2 = \begin{vmatrix} a_1 & a_3 \\ a_0 & a_2 \end{vmatrix}, \quad Q_m = \begin{vmatrix} a_1 & a_3 & a_5 & \cdots & a_{2m-1} \\ a_0 & a_2 & a_4 & \cdots & a_{2m-2} \\ \vdots & \vdots & \vdots & \vdots & \vdots \\ 0 & 0 & 0 & \cdots & a_m \end{vmatrix}$$

Chebotarev's Theorem [37]. *Assume that the functions $v(z)$ and $u(z)$ have no common zeros. Then quasi-polynomial (3.1) has no p-zeros if and only if*

$$Q_m > 0, \quad 1, 2, \ldots \tag{3.5}$$

Applications of this theorem are not effective because an infinite number of inequalities (3.5) must be verified.

3.2. Method of D Subdivision. Vyshnegradskii Diagrams

For determination of conditions under which quasi-polynomial (3.1) has no p-zeros, the method of D subdivision is widely applied [37, 59, 160(1), 160(2)]. The method of D subdivision is based on the fact that zeros of quasi-polynomial (3.1) are continuous functions of parameters a_{lj} and b_j. Construct the subdivision of the coefficient's space by hypersurfaces, the points of which are quasi-polynomials with at least one imaginary root. Such a procedure is called a D subdivision. For continuous variation of quasi-polynomial parameters the number of p-zeros may change only by passage of some zeros through an imaginary axis. Therefore points of every domain of D subdivision correspond to quasi-polynomials with the same number of p-zeros counted according to their multiplicities. Define now the number of p-zeros at least in one point of every domain of D subdivision of the corresponding quasi-polynomial. Then for all points of every domain the number of quasi-polynomial p-zeros will be just the same. D subdivision has a visual form when the quasi-polynomial depends on two parameters. Substitute $z = i\omega$ in characteristic equation (3.1), linearly depending on two parameters a and b. Equating to zero the real and imaginary parts, we have

$$\begin{aligned} U(\omega) &= aP_1(\omega) + bQ_1(\omega) + R_1(\omega) = 0 \\ V(\omega) &= aP_2(\omega) + bQ_2(\omega) + R_2(\omega) = 0 \end{aligned} \tag{3.6}$$

where $P_1(\omega), \ldots, R_2(\omega)$ are some continuous functions. From Eqs. (3.6) we obtain

$$a = \Delta_1/\Delta, \quad b = \Delta_2/\Delta, \quad \Delta = P_1 Q_2 - P_2 Q_1 \tag{3.7}$$

The formulae (3.7) determine the point $a_1 = a(\omega_1)$, $b = b(\omega_1)$ of D curve for $\omega = \omega_1$. If ω varies from $-\infty$ to ∞ we obtain some curves. Besides these

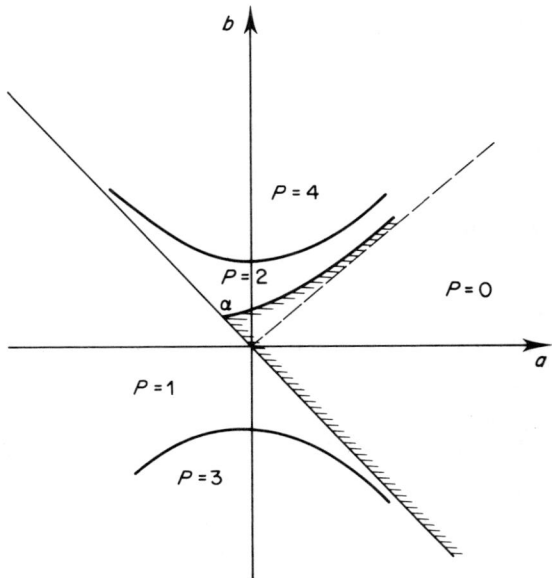

Fig. 2.1. D subdivision for quasi-polynomial (3.9).

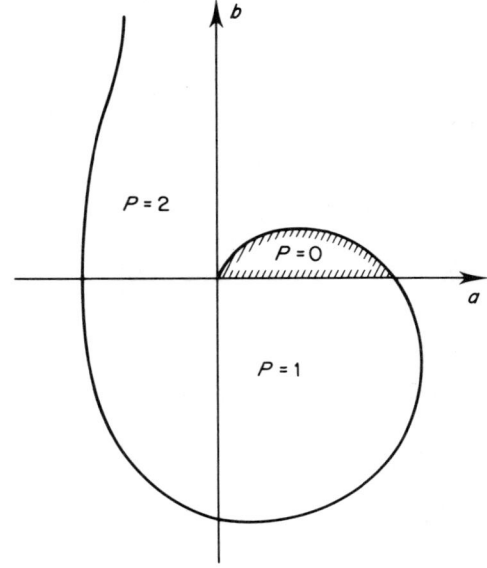

Fig. 2.2. Stability domain for the equation $\ddot{x}(t) + ax(t-1) + bx(t-1) = 0$.

§3. Methods of Stability Investigation of Linear Autonomous RFDEs

curves, D subdivision contains singular straight lines such that either $\Delta = 0$, $\Delta_1 = 0$ or $\Delta = 0$, $\Delta_2 = 0$ or $\Delta = \infty$, $\Delta_1 = \infty$. Consider for example, D subdivision of the equation

$$\dot{x}(t) + ax(t) + bx(t - h) = 0 \qquad (3.8)$$

Characteristic quasi-polynomial of Eq. (3.8)

$$D(z) = z + a + b\exp(-zh) \qquad (3.9)$$

has the root $z = 0$ for $a + b = 0$. Quasi-polynomial (3.9) has an imaginary root $z = i\omega$ if and only if $a + b\cos\omega h = 0$ and $y - b\sin\omega h = 0$. So the boundaries of D subdivision are singular straight line $a + b = 0$ and parametric curves $a = -(\omega\cos\omega h)/(\sin\omega h)$, $b = \omega/(\sin\omega h)$, $-\infty < \omega < \infty$. D subdivision for quasi-polynomial (3.9) is represented in Fig. 2.1. Quasi-polynomial (3.9) has no p-zeros for $a > 0$ and $b = 0$. Hence Eq. (3.8) is asymptotically stable in the domain shaded in Fig. 2.1. The number of p-zeros in other domains of D subdivision is indicated in Fig. 2.1. The stability domain for the equation $\ddot{x}(t) + a\dot{x}(t - 1) + bx(t - 1) = 0$ is represented in Fig. 2.2. D subdivision for the equation $\ddot{x}(t) + a\dot{x}(t - 1) + bx(t) = 0$ is given

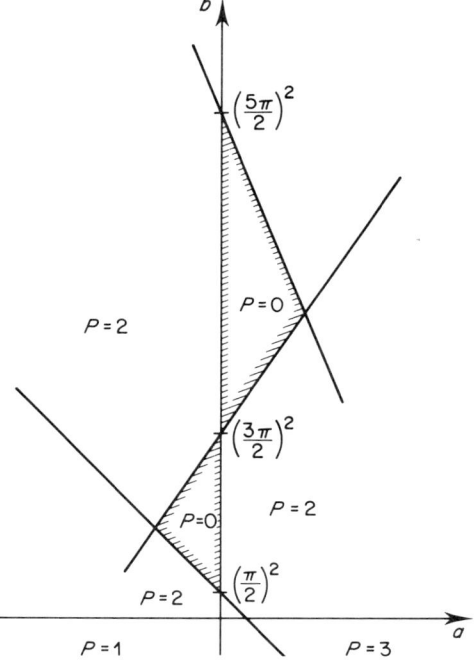

Fig. 2.3. D subdivision for the equation $\ddot{x}(t) + ax(t - 1) + bx(t) = 0$.

in Fig. 2.3. The stability domain of this equation consists of an infinite number of triangles adjoining the axis $a = 0$. The stability domain for the equation $\ddot{x}(t) + a\dot{x}(t) + bx(t - 1) = 0$ is represented in Fig. 2.4.

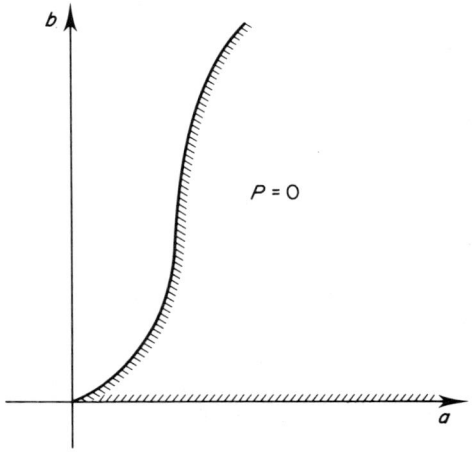

Fig. 2.4. Stability domain for the equation $\ddot{x}(t) + ax(t) + bx(t - 1) = 0$.

The formulated results enable us to find the stability domains of the simplest feedback systems from Table 1.2. In the theory of automatic control, these stability domains are often called *Vyshnegradskii diagrams* [219(2), 241(2)]. Consider for example the system with P controller

$$\dot{x}(t) + T^{-1}x(t) + KT^{-1}x(t - \tau) = KT^{-1}u(t - \tau) \qquad (3.10)$$

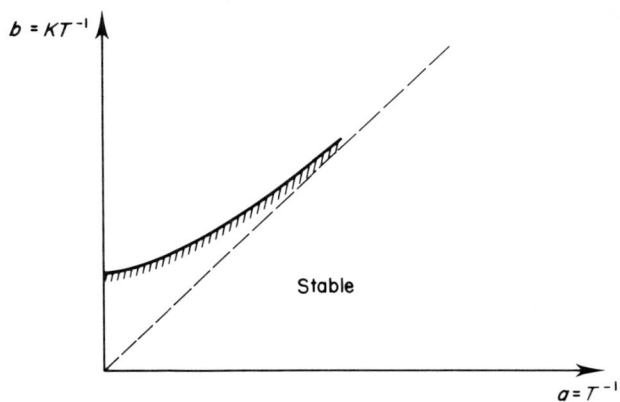

Fig. 2.5. Vyshnegradskii diagram for system (3.10) in parameters $a = T^{-1}, b = KT^{-1}$.

§3. Methods of Stability Investigation of Linear Autonomous RFDEs

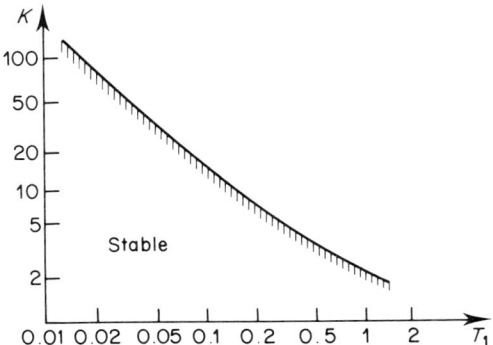

Fig. 2.6. Vyshnegradskii diagram for system (3.10) in parameters $T_1 = T^{-1}$, K.

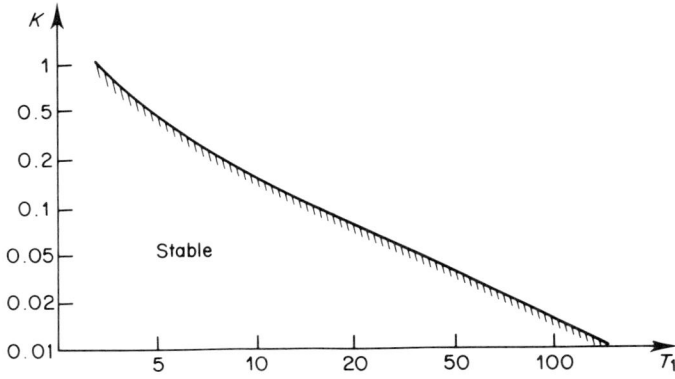

Fig. 2.7. Vyshnegradskii diagram for a system closed by I controller.

The Vyshnegradskii diagram of system (3.10) for $a = T^{-1}$ and $b = KT^{-1}$ is given in Fig. 2.1. But physically sensible feedback systems are possible only for $K > 0$ and $T > 0$. Consequently, the system (3.10) stability domain is less than in Fig. 2.1. The Vyshnegradskii diagram of system (3.10) in parameters $a = T^{-1}$ and $b = KT^{-1}$ is represented in Fig. 2.5. Logarithmic scales are used in Fig. 2.6, 2.7 and 2.8. The Vyshnegradskii diagram for system (1.6) with an I controller is represented in Fig. 2.7 and with a PI controller in Fig. 2.8. For other diagrams for single-circuit feedback systems see Gorecki [68].

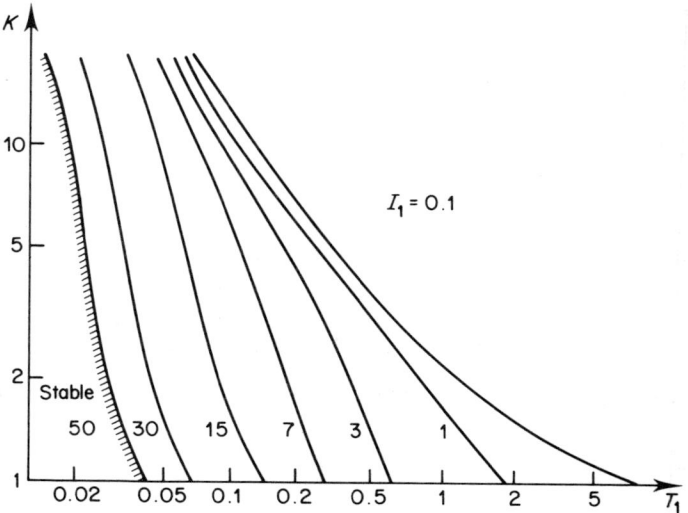

Fig. 2.8. Vyshnegradskii diagram for a system closed by a PI controller.

3.3. Michailov, Nyquist and Integral Frequency Criteria

Frequency methods are used in the theory of automatic control for investigation of the stability of autonomous systems. For RFDEs, frequency methods were developed by Kabakov [93] and Tzypkin [232(1)–232(3)]. These methods are based on the argument principle from the complex analysis [123]. The Michailov and Nyquist criteria are the most frequently used. Consider the characteristic functions (2.11)

$$\Delta(z) = \det\left[Iz - \int_0^\infty e^{-zs}\, dK(s) \right] = z^n + f_1(z)z^{n-1} + \cdots + f_n(z) \quad (3.11)$$

Here all the functions $f_j(z)$ are bounded and the function $\Delta(z)$ is analytical in the half-plane $\operatorname{Re} z \geq -\gamma$ for some $\gamma > 0$. Substituting $z = i\omega$ in (3.11), we get

$$\Delta(i\omega) = U(\omega) + iV(\omega) \quad (3.12)$$

The hodograph of function (3.12) in the complex plane is called a Michailov one.

Michailov Criterion. *For asymptotic stability of linear nth order equation with characteristic function (3.11), it is necessary and sufficient that variation of the function $\Delta(i\omega)$ will be equal to $n\pi/2$ when ω varies from 0 to ∞; i.e.,*

$$\left. \arg \Delta(i\omega) \right|_0^\infty = n\pi/2$$

§3. Methods of Stability Investigation of Linear Autonomous RFDEs

The Michailov criterion for RFDEs was proved by Kabakov [93] and Tzypkin [232(1)]. But for RFDEs, *Michailov hodographs* are more complicated than for ordinary differential equations. Typical Michailov hodographs for stable systems without delays are represented in Fig. 2.9. Michailov hodographs for

$$D(z) = 2z^2 + 0.5z + 2 + \exp(+z\tau) \tag{3.13}$$

are shown in Fig. 2.10 for $\tau = 1, 5, 10$. System (3.13) is unstable for $\tau = 1, 10$ and it is stable if $\tau = 5$.

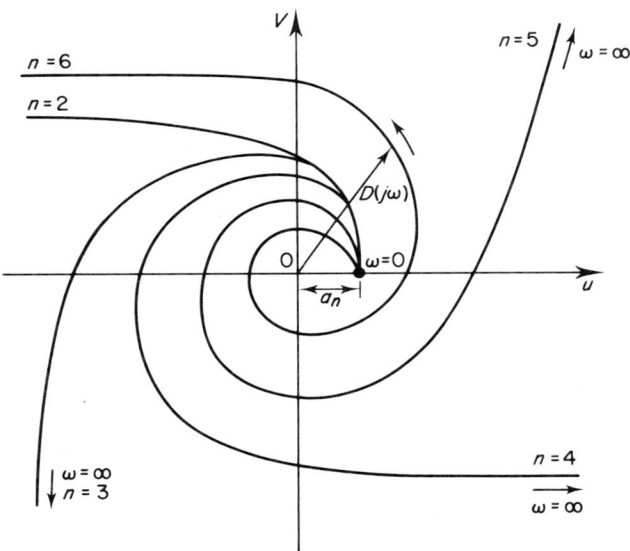

Fig. 2.9. Typical Michailov hodographs for nth-order systems without delay.

Let the transfer function of a linear RFDE be equal to $W(s)$. The function $W(i\omega)$ is called a *frequency response* of this system.

Nyquist Criterion. *Let the open-loop system be stable. Then for asymptotic stability of the closed-loop system it is necessary and sufficient that frequency response of the open-loop system does not envelope the point* (-1) *(see Fig. 2.11).*

The use of Michailov and Nyquist criteria for RFDEs is difficult because of the complexity of the corresponding curves.

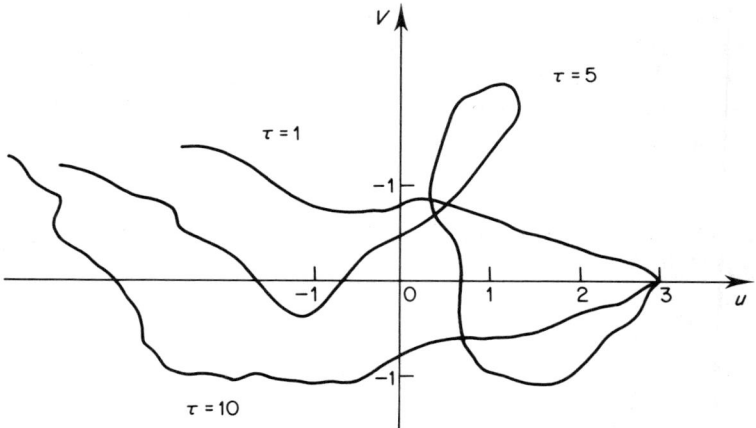

Fig. 2.10. Michailov hodographs for quasi-polynomials (3.13).

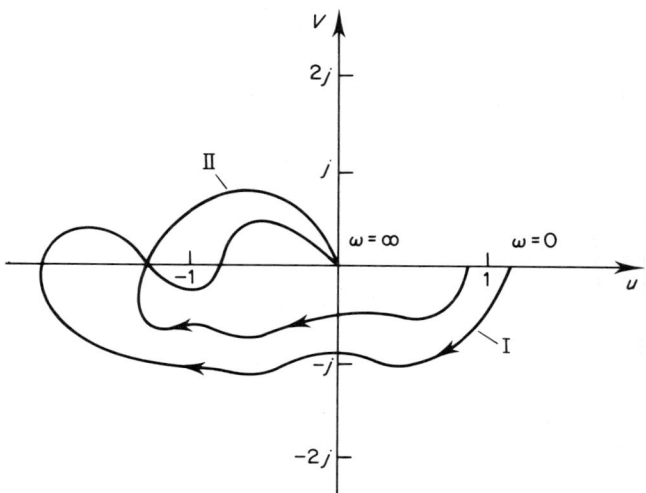

Fig. 2.11. Nyquist criterion: I, stable system; II, unstable system.

Integral criteria are more effective. Define the function of logarithmic derivative $R(\omega)$ for a system with characteristic function $\Delta(z)$

$$R(\omega) = \operatorname{Re} \frac{\Delta'(z)}{\Delta(z)} = \frac{d}{d\omega} [\arg F(i\omega)]$$

$$= \frac{U(\omega)V'(\omega) - V(\omega)U'(\omega)}{U^2(\omega) + V^2(\omega)} \tag{3.14}$$

§3. Methods of Stability Investigation of Linear Autonomous RFDEs

where $\Delta(i\omega) = U(\omega) + iV(\omega)$. From the argument principle the integral criterion of stability follows.

Integral Criterion of Stability [144]. *For asymptotic stability of RFDE with characteristic function $\Delta(z)$, it is necessary and sufficient that*

$$\int_0^\infty R(\omega)\, d\omega = \frac{n\pi}{2} \tag{3.15}$$

Here $R(\omega)$ is defined by formula (3.14)

The integral stability criterion is more convenient for computer calculations. Choose a number s such that

$$\int_s^\infty R(\omega)\, d\omega < 1$$

Then instead of condition (3.15) it is sufficient to verify the inequality

$$\int_s^\infty R(\omega)\, d\omega > \frac{(n-1)\pi}{2}$$

Consider, for example, the system with characteristic function

$$\Delta(z) = 0.1z^2 + 0.3z + 0.5 + (0.1z + 0.2)\exp(-z\tau_1)$$
$$+ (0.2z + 0.3)\exp(-z\tau_2)$$

We get

$$U = -0.1\omega^2 + 0.5 + 0.1\omega \sin \tau_1\omega + 0.2 \cos \tau_1\omega$$
$$+ 0.2\omega \sin \tau_2\omega + 0.3 \cos \tau_2\omega$$
$$V = 0.3\omega + 0.1\omega \cos \tau_1\omega - 0.2 \sin \tau_1\omega + 0.2\omega \cos \tau_2\omega$$
$$- 0.3 \sin \tau_2\omega$$

The function $R(\omega)$ from (3.14) is represented in Fig. 2.12 for $\tau_1 = 3.0, \tau_2 = 1.5$ and in Fig. 2.13 for $\tau_1 = 2.5, \tau_2 = 1.7$. Computing the integrals we have

$$\int_0^{20} R(\omega)\, d\omega = 3.0455; \qquad \tau_1 = 3.0, \quad \tau_2 = 1.5$$
$$\int_0^{20} R(\omega)\, d\omega = -1.0834 < \pi; \qquad \tau_1 = 2.5, \quad \tau_2 = 1.7 \tag{3.16}$$

At the first case the system is stable and it is unstable at the second one. Remark that studying the function $R(\omega)$ we obtain some estimations of the transient regimes (see Melkumjan [144] and §4). The *wave criterion* for RFDEs is developed in Lekus and Rovinskii [124].

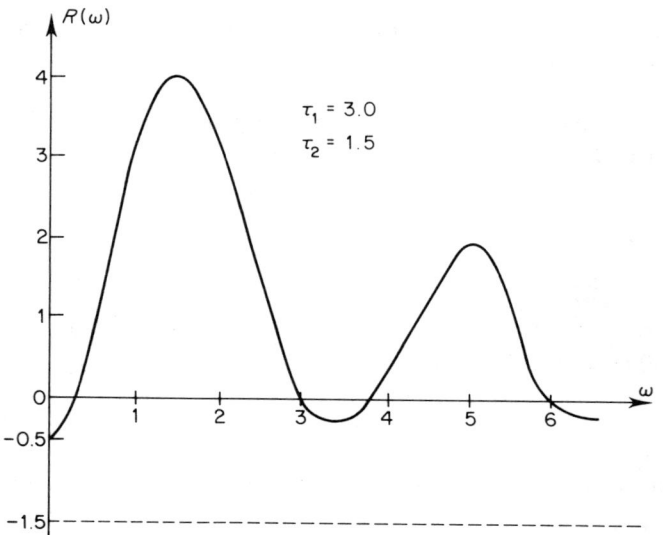

Fig. 2.12. Logarithmic derivative $R(\omega)$ of stable system.

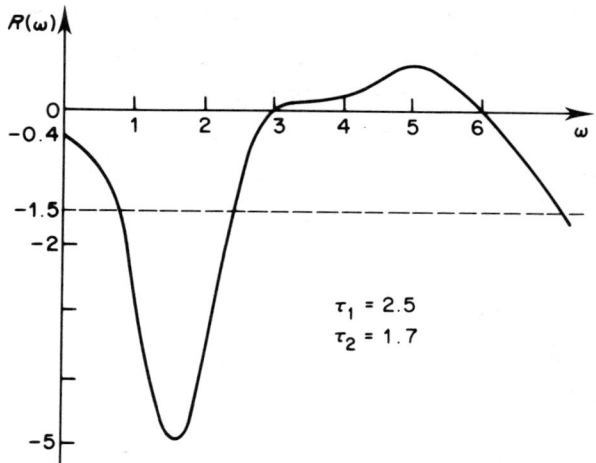

Fig. 2.13. Logarithmic derivative $R(\omega)$ of unstable system.

3.4. Stability for Arbitrary Delays. Tsypkin Criterion

In some problems it is interesting to have the conditions under which an RFDE is stable for arbitrary delays. Consider an open-loop system with transfer function $W(s) = [R_{n-1}(s)/Q_n(s)] \exp(-s\tau)$, where $R_{n-1}(s)$ and $Q_n(s)$

§3. Methods of Stability Investigation of Linear Autonomous RFDEs

are polynomials of degrees $(n - 1)$ and n. The closed-loop system (Fig. 2.14) has the following transfer function

$$R_{n-1}(s) \exp(-s\tau)/[R_{n-1}(s) \exp(-s\tau) + Q_n(s)] \tag{3.17}$$

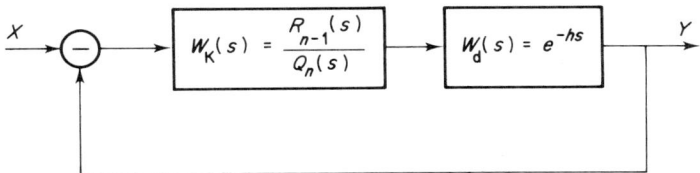

Fig. 2.14. System with separated delay unit.

Tsypkin Criterion [232(2)]. *Let $Q_n(s)$ be a stable polynomial. Then the closed-loop system (3.17) is stable for arbitrary delays $0 \le \tau < \infty$ if and only if*

$$|Q_n(i\omega)| > |R_{n-1}(i\omega)|, \qquad -\infty < \omega < \infty \tag{3.18}$$

For the proof let us remark that the frequency response of system (3.17) may be constructed in two steps. First, one constructs the curve $R_{n-1}(i\omega)/Q_n(i\omega)$. By virtue of (3.18) this curve is situated entirely within the unit circle. But the action of the factor $\exp(-i\tau\omega)$ is reduced to the rotation of the vector $R_{n-1}(i\omega)/Q_n(i\omega)$ on the angle $(-\tau\omega)$. So the frequency response curve of system (3.16) is situated entirely within the unit circle also and cannot envelope the point (-1). By the Nyquist criterion, the closed-loop system (3.17) is stable.

Now let $|Q_n(i\omega)_{cr})| = |R_{n-1}(i\omega_{cr})|$ for some ω_{cr}. Then it is always possible to choose the rotation angle $(-\tau_{cr}\omega_{cr})$ such that the frequency response passes through the point (-1). In addition, system (3.17) is stable for $0 < \tau < \tau_{cr}$ and unstable for $\tau_{cr} < \tau$.

In El'sqol'tz [59], and in volume 7 of [226], the Tsypkin criterion is generalized on the systems with transfer function

$$W(s) = \left[\sum_{j=0}^{m} R_j(s) \exp(-s\tau_j) \right] \bigg/ Q(s) \tag{3.19}$$

$$Q(s) = s^n + a_1 s^{h-1} + \cdots + a_n, \qquad R_j(s) = b_{0j}s^n + b_{1j}s^{n-1} + \cdots + b_{nj}$$

Assume that $|b_{01}| + \cdots + |b_{0m}| < 1$ and that the polynomial $Q(s)$ is stable. Then for asymptotic stability of the closed-loop system (3.19) for arbitrary delays τ_i it is necessary and sufficient that

$$\sum_{j=0}^{m} |R_j(i\omega)| < |Q(i\omega)|$$

The equivalent form of this statement is the following: Under formulated conditions, the quasi-polynomial

$$D(s) = Q(z) + \sum_{j=0}^{m} R_j(z) \exp(-\tau_j z)$$

has no p-zeros for arbitrary delays $\tau_j \geq 0$.

The problem of stability for arbitrary delays of the system

$$\dot{x}(t) = Ax(t) + Bx(t-\tau), \qquad x(t) \in R_n \qquad (3.20)$$

is investigated in Lewis and Anderson [126], and Repin [183(1)]. It is proved that if the matrix A is stable and all the roots λ_j of the equation

$$\det \begin{bmatrix} \lambda A - B & -\lambda \mu E \\ -\lambda \mu E & \lambda A - B \end{bmatrix} = 0$$

lie inside the unit circle for all real μ, then system (3.20) is stable for all delays $\tau \geq 0$ [183(1)].

3.5. General Function (2.11) [108(5)]

Consider the scalar equation

$$x^{(n)}(t) = \sum_{j=0}^{n-1} \int_0^\infty x^{(j)}(t-s) \, dk_j(s), \qquad t \geq 0$$

Assume that the integral

$$\alpha_{lj} = \int_0^\infty s^l |dk_j(s)|, \qquad \beta_{lj} = \int_0^\infty s^l \, dk_j(s)$$

are absolutely convergent. Formulate some conditions for the absence of p-zeros for the characteristic function

$$\Delta_n(z) = z^n - \sum_{j=0}^{n-1} z^j \int_0^\infty \exp(-zs) \, dk_j(s)$$

in the cases $n = 1, 2$.

(a) Let the conditions $\beta_{00} < 0$, $\alpha_{10} < 1$ be fulfilled. Then equation $\Delta_1(z) = 0$ has no p-zeros.

(b) Let $k_0(s)$ be nonincreasing and $k_0(s) = \text{const}$ for $s \geq h \geq 0$. If $\beta_{00} < 0$ and $h\alpha_{00} < \pi/2$, then the equation $\Delta_1(z) = 0$ has no p-zeros. The estimation $h\alpha_{00} < \pi/2$ is exact (see Fig. 2.1).

REMARK 3.1. The condition $\beta_{00} < 0$ is necessary for the absence of p-zeros for the equation $\Delta_1(z) = 0$. Otherwise, there exists a nonnegative real root of this equation.

REMARK 3.2. Let the function $k_0(s)$ have a jump $a_0 < 0$ in the point $s = 0$. Then the equation $\Delta_1(z) = 0$ has no p-zeros if

$$\int_{+0}^{\infty} |dk_0(s)| < -a_0 \tag{3.21}$$

If, in addition, the function $k_0(s)$ is nondecreasing for $s > 0$, then by virtue of Remark 3.1 condition (3.21) will be necessary.

(c) Let the function $k_1(s)$ have a jump $a_0 < 0$ in the point $s = 0$ and

$$\alpha_{11} < 2, \qquad \beta_{00} < 0, \qquad \int_{+0}^{\infty} |dk_1(s)| + \alpha_{10} < -a_0$$

Then the function $\Delta_2(z)$ has no p-zeros.

(d) Assume that $\beta_{00} < 0$, $\alpha_{11} + \alpha_{20}/2 < 1$ and $\beta_{10} - \beta_{01} > (-2\beta_{00})^{1/2} > 0$. Then the function $\Delta_2(z)$ has no p-zeros.

For the proofs of statements (a)–(d) see Kolmanovskii [108(5)].
As an example consider the equation

$$\ddot{x}(t) = -ax(t) + bx(t-h), \qquad t \geq 0 \tag{3.22}$$

Because of (d), the trivial solution of Eq. (3.22) is asymptotically stable if $2a > 2b > a > 0$ and $0 < h < (b/(a+2b))^{1/2}$. This example shows that delay may stabilize the system. The system which is unstable for $b = 0$ becomes asymptotically stable for some $b > 0$.

§4. STABILIZATION SYSTEM OF THE TWO-REFLECTOR ANTENNA

The two-reflector (or Cassegrainian) antenna is used to form a narrow directional source of radio-frequency radiation. The two-reflector antenna consists of a big immovable reflector and of a little movable one, furnished with a control system. The big reflector is a hemispherical metal surface approximately 50 m in diameter, with the axis forming a certain angle with the horizon. The little reflector has a specially calculated complex surface with a radiator situated in its focus. The radiator is placed on a massive platform having a diameter of 7 m. To reach high guidance accuracy the control system includes a platform stabilization system, the precision of which is about 5″ to 6″. The antenna has to function outdoors continuously by day and night under considerable wind and temperature disturbances. Platform deflections are measured by optical pickups. The actuator of the stabilization system contains three screws which can move three points of the platform in the vertical direction. Hence the stabilization system is multidimensional, more exactly three-dimensional, including three separate channels for each screw servomechanism. The position errors in the screw servomechanism are measured by hydrostatic pickups which have considerable lags, until 1 to 2 sec.

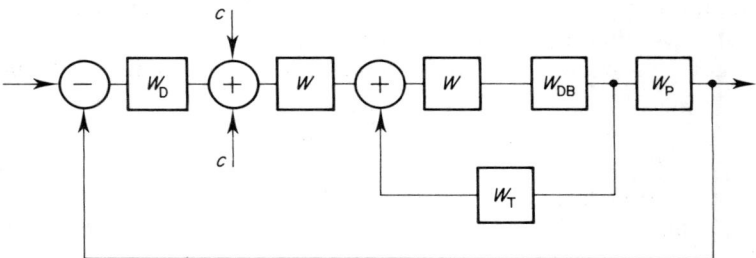

Fig. 2.15. Separate channel of screw servomechanism.

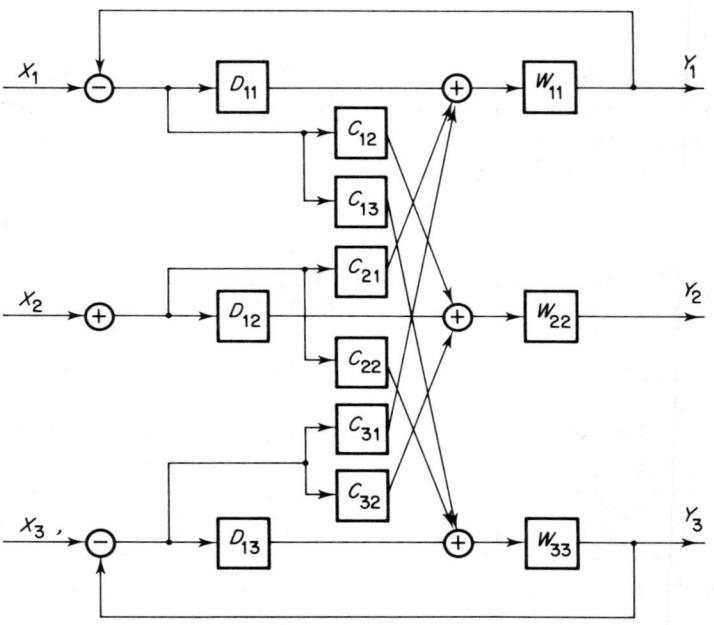

Fig. 2.16. Scheme of multidimensional stabilization system.

§4. Stabilization System of the Two-Reflector Antenna

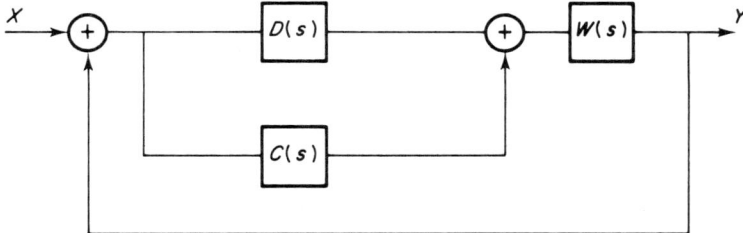

Fig. 2.17. Matrix structural scheme.

A separate channel of the screw servomechanism is represented in Fig. 2.15. Figure 2.16 shows a general scheme of a multidimensional stabilization system, and Fig. 2.17 gives a matrix structural scheme. The transfer-matrix function of the open-loop multidimensional system is

$$G(s) = W(s)[C(s) + D(s)] \tag{4.1}$$

$$W = \begin{bmatrix} W_{11} & 0 & 0 \\ 0 & W_{22} & 0 \\ 0 & 0 & W_{33} \end{bmatrix}, \quad D = \begin{bmatrix} D_{11} & 0 & 0 \\ 0 & D_{22} & 0 \\ 0 & 0 & D_{33} \end{bmatrix}, \quad C = \begin{bmatrix} 0 & C_{12} & C_{13} \\ C_{21} & 0 & C_{23} \\ C_{31} & C_{32} & 0 \end{bmatrix}$$

From the symmetry of the multidimensional system we have

$$\begin{aligned} W_{11} &= W_{22} = W_{33} = K_1 s^{-1}(T_e s + K_2)^{-1} \\ K_1 &= K_v K_{p0} K_e K_r, \quad K_2 = 1 + K_{p0} K_e K_f \\ D_{11} &= D_{22} = D_{33} = K_p \exp(-s\tau) \\ C_{ij} &= K_c \exp(-s\tau), \quad i,j = 1, 2, 3, \quad i \neq j \end{aligned} \tag{4.2}$$

Numerical values for parameters are received experimentally [144] and are as follows: for the engine $T = T_e = 0.11$ sec; $K_e = 17.58$ rad/V sec; reduction coefficient $K_r = 9 \times 10^{-4}$ mm/rad; pickup gain $K_p = 0.4$ V/mm; interconnection gain $K_c = 0.05$ V/mm; voltage gain $K_v = 250$; power gain $K_{p0} = 10$; local feedback gain $K_f = 0.03$ V sec/rad.

The transfer-matrix function of the closed-loop system is equal to $\Phi(s) = G(s)[I + G(s)]^{-1}$. The characteristic function of the closed-loop multidimensional system is

$$\begin{aligned} D(s) = \det[I + G(s)] &= P_0(s) + P_1(s) \exp(-s\tau) \\ &+ P_2(s) \exp(-2 s\tau) + P_3(s) \exp(-3 s\tau) \end{aligned} \tag{4.3}$$

where

$$P_0(s) = T^3s^6 + 3K_2T^2s^5 + 3TK_2^2s^4 + K_2^3s^3$$
$$P_1(s) = 3K_1T^2K_ps^4 + 6K_1K_2K_pTs^3 + 3K_1K_2^2K_ps^2$$
$$P_2(s) = 3K_1^2T(K_p^2 - K_c^2)s^2 + 3K_1^2K_2(K_p^2 - K_c^2)s$$
$$P_3(s) = K_1^3T(K_p^3 + 2K_c^3 - 3K_pK_c^2)$$

We use the integral criterion from §3 for stability investigation of a closed-loop multidimensional stabilization system. Define the function $R(\omega) = (U\dot{V} - V\dot{U})/(U^2 + V^2)$, $U = \text{Re } D(i\omega)$, $V = \text{Im } D(i\omega)$, $i^2 = -1$, where $D(i\omega)$ is expressed by relation (4.3).

After that we construct graphs of the functions $R(\omega)$ for different values of parameters such that voltage gain K_v, interconnection gain K_c and time lag τ. Use of these graphs permits us to determine the stability and some characteristics of transient processes. As it is shown by Melkumjan [144] the following estimates hold: transient time $t_t \leq a \max_\omega[R(\omega) + \tau/2]$, $a = \frac{3}{5}$; number of oscillations $N \leq \max_\omega[\omega R(\omega) + \omega\tau/2]$; time t_{max} of the first maximum of the transient process $t_{max} = \pi N^{-1}$; value of overshoot $\sigma \leq \exp[-t_{max}t_ta^{-1}]$. The family of graphs $R(\omega)$ for $\tau = 0.3$ sec, $K_v = 200$ and $K_f = 0.001, 0.02, 0.03$ V/sec rad is represented in Fig. 2.18. It is clear that a stabilization system is unstable if there is no local feedback, i.e., $K_f = 0$ V/sec rad.

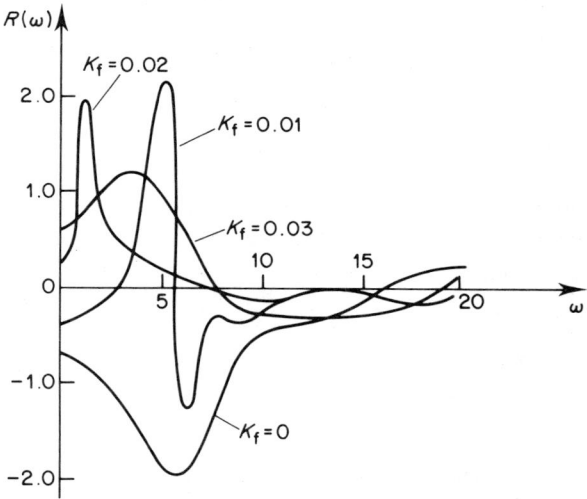

Fig. 2.18. Family of logarithmic derivatives $R(\omega)$ for different K_f.

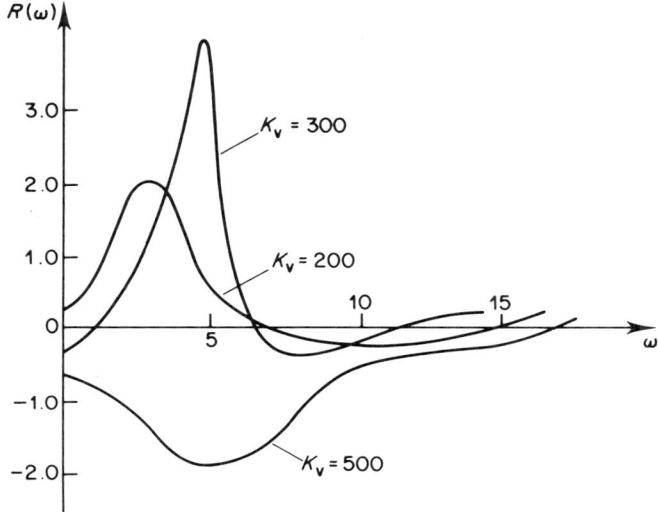

Fig. 2.19. Family of $R(\omega)$ for different K_v.

For $K_f \geq 0.02$ V/sec rad the system is stable. The number of oscillations N decreases with the increase of K_f. Fig. 2.19 shows a family of graphs $R(\omega)$ for $\tau = 0.3$ sec, $K_f = 0.02$ V/sec rad and $K_v = 200, 300, 500$. Together with the increase of K_v the transient time t_t also increases, system stability decreases and finally for $K_v = 500$ the system becomes unstable. For $K_1 = 15$, $K_2 = 6.3$, $T = 0.11$ one can evaluate critical lag τ_{cr}, which is equal to 0.5 sec. If $0 \leq \tau \leq \tau_{cr}$, then the system is stable; for $\tau > \tau_{cr}$, the system is unstable. These results have been confirmed by experiments [144].

§5. LIAPUNOV DIRECT METHOD FOR EQUATIONS WITH DELAY

Stability theory of RFDEs is one of the important directions in the investigations of these equations. Stability theory is developed in several directions. First, stability of concrete types of equations are studied (for example, linear autonomous equations, linear periodic equations, quasi-linear equations and so on). Further, general methods similar to the Liapunov direct method are created.

In this section some fundamental ideas and results relating to the Liapunov direct method for RFDEs are presented.

5.1. Application of Liapunov Functions. Razumikhin-Type Theorems

Consider the initial-value problem for the RFDE

$$\dot{x}(t) = f(t, x_t), \quad t \geq t_0, \quad x_t(\theta) = x(t + \theta)$$
$$x_{t_0}(\theta) = \varphi(\theta), \quad -h \leq \theta \leq 0 \tag{5.1}$$

Assume as in §1 that $\varphi(\theta) \in C[-h, 0]$, the operator $f(t, \varphi)$, $f: R_1 \times Q_H \to R_n$ is continuous and Lipschitzian in $\varphi \in Q_H$ and

$$f(t, 0) = 0 \tag{5.2}$$

Investigate the stability of the trivial solution $x(t) = 0$ of Eq. (5.1). This is possible by the method of the Liapunov function [3, 65, 180(1), 180(2), 197]. Formulate a Razumikhin-type theorem.

Theorem 5.1. *Let there exist, for Eq. (5.1), the continuous positive-definite function $V(t, x)$, $V: R_1 \times R_n \to R_n$, whose derivative computed along the trajectories of Eq. (5.1) is nonpositive for any solution $x(t)$ satisfying the inequality*

$$V(s, x(s)) \leq V(t, x(t)), \quad s \leq t, \quad t \geq t_0 \tag{5.3}$$

Then the trivial solution of (5.1) is stable. If, in addition the function $V(t, x)$ has an infinitesimal upper limit and its derivative $\dot{V}$ is negative-definite along any solution $x(t)$ of Eq. (5.1) such that

$$V(s, x(s)) \leq f(V(t, x(t)), \quad s \leq t, \quad t \geq t_0 \tag{5.4}$$

where f is a continuous function and also $f(u) > u$ for $u > 0$. Then the trivial solution of (5.1) is uniformly asymptotically stable.

For the proof of Theorem 5.1 see Hale [81(1)], and Razumikhin [180(1), 180(2)].

5.2. Method of Liapunov-Krasovskii Functionals for Equations with Bounded Delay

Krasovskii has proposed to use for stability investigation the functionals defined on $C[-h, 0]$. Theorems 5.2 and 5.3 are proved by Krasovskii [114(1)–114(5)]. Let $V: R_1 \times Q_H \to R_1$ be some continuous functional such that

$$V(t, 0) = 0 \tag{5.5}$$

Denote by $\omega_i(r)$, $r \geq 0$, some scalar, continuous, nondecreasing functions

§5. Liapunov Direct Method for Equations with Delay

such that $\omega_i(0) = 0$ and $\omega_i(r) > 0$ for $r > 0$. The functional $V(t, \varphi)$ is *positive-definite* if there exists a function ω_1 such that

$$V(t, \varphi) \geq \omega_1(|\varphi(0)|), \qquad \varphi \in Q_H, \quad t \in R_1 \tag{5.6}$$

The functional $V(t, \varphi)$ has an *infinitesimal upper limit* if there exists a function ω_2 such that

$$V(t, \varphi) \leq \omega_2(\|\varphi(\theta)\|) \tag{5.7}$$

Substitute the trajectory $x_t(\theta)$, $-h \leq \theta \leq 0$ of (5.1), instead of $\varphi(\theta)$, into the functional $V(t, \varphi)$ and denote $V(t)$ by $V(t, x_t)$. The *derivative* $\dot{V}$ *of the functional* $V(t, x_t)$ *computed along the trajectories of* (5.1) is equal by definition to

$$\dot{V} = \lim_{\Delta t \to +0} \sup[V(t + \Delta t) - V(t)]/(\Delta t)$$

In other words the derivative $\dot{V}$ is the right-hand upper derivative of the functional V on the trajectories of (5.1). The derivative $\dot{V}$ is called *negative-definite* if there exists such a function ω_3 that $\dot{V} \leq -\omega_3(|x(t)|)$.

Theorem 5.2. *Let there exist continuous positive-definite functional* $V(t, \varphi)$ *such that* $\dot{V} \leq 0$. *Then the trivial solution of Eq.* (5.1) *is stable.*

Proof. Assume any $\varepsilon \in (0, H)$. By virtue of (5.5) we can take $\delta(\varepsilon, t_0)$ such that

$$\max_{\|\varphi(\theta)\| \leq \delta(\varepsilon, t_0)} V(t_0, \varphi(\theta)) \leq \omega_1(\varepsilon) \tag{5.8}$$

From condition $\dot{V} \leq 0$ it follows that V is nonincreasing along the trajectories of (5.1). Hence using (5.6) and (5.8) we have

$$\omega_1(|x(t)|) \leq V(t, x_t) \leq V(t_0, x_{t_0})$$
$$= V(t_0, \varphi) \leq \omega_1(\varepsilon), \qquad t \geq t_0$$

The monotonicity of ω_1 implies that $|x(t)| \leq \varepsilon$, $t \geq t_0$. Theorem 5.2 is proved.

REMARK. If, in addition to the assumptions of Theorem 5.2, the functional $V(t, \varphi)$ satisfies (5.7), then the trivial solution is *uniformly stable*.

Theorem 5.3. *Let there exist a continuous functional* $V(t, \varphi)$ *such that*

$$\omega_1(|\varphi(0)|) \leq V(t, \varphi) \leq \omega_2(\|\varphi(\theta)\|), \qquad \dot{V} \leq -\omega_3(|x(t)|) \tag{5.9}$$

Then the trivial solution of (5.1) *is uniformly asymptotically stable. Conversely, if the trivial solution of* (5.1) *is uniformly asymptotically stable then there exists* a continuous functional $V(t, \varphi)$ satisfying (5.9) and a local Lipschitz condition in φ.

For the proof of Theorem 5.3 see Krasovskii [114(5)]

EXAMPLE 5.1. Consider a scalar equation

$$\dot{x}(t) = -ax(t) + b(t)x(t-h), \qquad a > 0, \quad h > 0, \quad t \geq t_0 \qquad (5.10)$$

where $b(t)$ is a continuous function. Introduce the functional

$$V(t, x_t) = x^2(t) + a \int_{t-h}^{t} x^2(s)\, ds$$

The derivative $\dot{V}$ is such that

$$\dot{V} = 2x(t)[-ax(t) + b(t)x(t-h)] + ax^2(t) - ax^2(t-h)$$
$$\leq -v[x^2(t) + x^2(t-h)], \qquad v = a - \sup_{t \geq t_0} |b(t)|$$

By Theorem 5.3 the trivial solution of Eq. (5.10) is uniformly asymptotically stable if $v > 0$.

5.3. Equations with Unbounded Delay

Consider an RFDE with unbounded delay

$$\begin{aligned}\dot{x}(t) &= f(t, x_t), & t &\geq t_0 \\ x_{t_0} &= x(t_0 + \theta) = \varphi(\theta), & -\infty &< \theta \leq 0\end{aligned} \qquad (5.11)$$

where the initial function $\varphi(\theta)$ is continuous. As in §1 denote M a metric space of continuous functions $\varphi(\theta)$ $\varphi: (\infty, 0] \to R_n$ with a metric ρ. Denote Q_H a sphere in the space M, i.e., $Q_H = \{\varphi \in M, \rho(\varphi, 0) \leq H\}$. Let the map $f: [0, \infty) \times M \to R_n$ satisfy conditions sufficient for existence and uniqueness of the problem (5.11) solution. Assume also

$$f(t, 0) = 0, |f(t, \varphi)| \leq F_0, \qquad t \geq t_0, \quad \varphi \in Q_H \qquad (5.12)$$

Notice that Theorems 5.2, 5.3 are not directly applied to Eq. (5.11) with unbounded delay.

Theorem 5.4. *Let there exist a continuous functional $V(t, \varphi)$, $V: [0, \infty) \times Q_H \to R_1$ such that*

$$\omega_1(|\varphi(0)|) \leq V(t, \varphi) \leq \omega_2(\rho(\varphi, 0)) \qquad (5.13)$$

$$\dot{V} \leq -\omega_3(|x(t)|) \qquad (5.14)$$

Then the trivial solution of Eq. (5.11) is asymptotically stable.

§5. Liapunov Direct Method for Equations with Delay

Proof. Show first that the solution $x(t) \equiv 0$ of (5.11) is stable. Assume any $\varepsilon > 0$. Choose $\delta > 0$ such that $\omega_1(\varepsilon) = \omega_2(\delta)$. Using (5.13) and (5.14), we obtain for any $\varphi \in Q_\delta$ that

$$\omega_1(|x(t)|) \leq V(t, x_t) \leq V(t_0, \varphi) \leq \omega_2(\rho(\varphi, 0))$$
$$\leq \omega_2(\delta) = \omega_1(\varepsilon) \tag{5.15}$$

The monotonicity of ω_1 implies that $|x(t, t_0, \varphi(\theta))| \leq \varepsilon$ for $\varphi(\theta) \in Q_\delta$. Persuade oneself that $x(t, t_0, \varphi(\theta)) \to 0$ for $t \to \infty$ and for any $\varphi(\theta) \in Q_{H_1}$. Here H_1 is such that $\omega_1(H) = \omega_2(H_1)$. Reasoning as in (5.15) we get that $|x(t, t_0, \varphi(\theta))| \leq H$ for $t \geq 0$ and $\varphi \in Q_{H_1}$. Assume now that the solution $x(t, t_0, \varphi(\theta))$ does not tend to zero for some $\varphi(\theta) \in Q_{H_1}$. Then for some $\varepsilon > 0$ there exists a sequence $\{t_i\}$ such that $|x(t_i, t_0, \varphi(\theta))| \geq \varepsilon$ and $t_i \to \infty$. By (5.12) the derivative $\dot{x}(t, t_0, \varphi(\theta))$ is bounded: $|\dot{x}(t)| \leq F_0$. Therefore, for all s such that $|s| \leq \Delta = \varepsilon/(2F_0)$ we have $|x(t_i + s, t_0, \varphi(\theta))| \geq \varepsilon/2$. From (5.14) it follows $\dot{V} \leq -\omega_3(\varepsilon/2) = v < 0$ for $|s| \leq \Delta$. So

$$V(t) - V(t_0) = \int_{t_0}^t \dot{V}\, dt \geq \sum_{t_i \leq t} \int_{t_i - \Delta}^{t_i + \Delta} \dot{V}(t)\, dt = 2v\Delta\, N(t)$$

where $N(t)$ is a number of t_i such that $t_i \leq t$. But $N(t) \to \infty$ for $t \to \infty$. Consequently, $V(t) - V(t_0) \to -\infty$ for $t \to \infty$. This contradiction with (5.13) shows that $x(t, t_0, \varphi(\theta)) \to 0$ for $t \to \infty$ and $\varphi \in Q_{H_1}$. Theorem 5.4 is proved.

5.4. Stability in the First Approximation and under Steady Acting Perturbations

Consider the equation

$$\dot{x}(t) = L(x_t) + R(t, x_t), \qquad x_t = x(t + \theta), \qquad -h \leq \theta \leq 0 \tag{5.16}$$

Here $L(\varphi): C[-h, 0] \to R_n$ is a continuous linear mapping and $R(t, \varphi): R_1 \times C[-h, 0] \to R_n$ is a continuous mapping. Along with (5.16) consider the *equation of first approximation*

$$\dot{x}(t) = L(x_t) = \int_{-h}^0 [dK(\theta)]x(t + \theta), \qquad -h \leq \theta \leq 0 \tag{5.17}$$

where $K(\theta)$ is a matrix whose elements are functions with bounded variations. The characteristic function of Eq. (5.17) is $D(z)$,

$$D(z) = \det\left[\int_h^0 [dK(\theta)]e^{z\theta} - Iz\right] \tag{5.18}$$

Assume that

$$|R(t, \varphi)| \leq \beta\|\varphi(\theta)\| \tag{5.19}$$

Theorem 5.5. *Let zeros z_i of function (5.18) satisfy the condition* Re $z_i \leq -v < 0$. *Then there exists a constant $\beta > 0$ such that the trivial solution of (5.16) is uniformly asymptotically stable for any mapping $R(t, \varphi)$ complying with (5.19)* [114(5)].

Requirement (5.19) in Theorem 5.5 may be weakened until the following: there is a number $T > 0$ and a continuous function $\psi(s) \geq 0$ such that

$$|R(t, \varphi)| \leq \psi(t)\|\varphi(\theta)\|, \qquad \frac{1}{T}\int_t^{t+T} \psi(s)\,ds \leq \beta$$

For nonautonomous or nonlinear equations of first approximation, complete results are not derived, although there are some theorems in this direction [79(4), 81(4)]. In particular, if the trivial solution of the first approximation equation is exponentially stable, then the trivial solution of the full equation is asymptotically stable [108(5), 114(5)].

Let us investigate stability under steady-acting disturbances. Along with Eq. (5.1) consider the disturbed systems

$$\dot{x}(t) = f(t, x_t) + R(t, x_t) \qquad (5.20)$$

The trivial solution of (5.1) is called *stable under steady-acting disturbances* if for any $\varepsilon > 0$ there exist $\delta > 0$ and $\eta > 0$ such that all solutions $x(t, t_0, \varphi)$ satisfy the inequality $|x(t, t_0, \varphi)| \leq \varepsilon$ for $\|\varphi\| \leq \delta$ and $|R(t, \psi)| \leq \eta$, where $\|\psi\| \leq \varepsilon$. Krasovskii [114(5)] has proved that the trivial solution of (5.1) is stable under steady-acting disturbances if it is uniformly asymptotically stable. It is possible to replace the condition of smallness of the disturbance's norm by the assumption of its smallness in the average [63, 128].

5.5. Case of a Nonpositive Derivative

Construction of Liapunov–Krasovskii functionals satisfying the conditions of Theorems 5.2 and 5.3 involve some difficulties. Sometimes it is easier to construct the functionals with a nonpositive derivative. The corresponding theorems follow.

First we formulate a statement which extends the *Barbashin–Krasovskii* theorem. Consider an autonomous RFDE of the form (5.1):

$$\dot{x}(t) = f(x_t), \qquad t \geq 0, \ x_t = x(\cdot + \theta), \qquad -h \leq \theta \leq 0 \qquad (5.21)$$

By definition, the element $\psi \in C[-h, 0]$ belongs to an ω-*limit set* corresponding to the initial function φ if $x(t, 0, \varphi)$ is defined on $[0, \infty)$ and there exists a sequence $\{t_n\}$, $t_n \to \infty$ for $n \to \infty$ such that $\|x_{t_n}(\varphi) - \psi\| \to 0$, $n \to \infty$, and $x_{t_n}(\varphi) = x(t_n + \theta, 0, \varphi)$. The set $Q \subseteq C[-h, 0]$ is called *invariant* if $x_t(\varphi) \in Q$, $t \in [0, \infty)$ for any $\varphi \in Q$.

It is possible to prove that the ω-limit set corresponding to any initial function is situated in the invariant set of Eq. (5.21).

Theorem 5.6 [81(1), 81(4)]. *Let there exist the continuous functional $V(\varphi): C[-h, 0] \to R_1$ such that*

$$\omega_1[|\varphi(0)|] \leq V(\varphi) \leq \omega_2[\|\varphi(\theta)\|]$$

$$\omega_1(t) \to \infty, \quad t \to \infty \quad \text{and} \quad \dot{V}(x_t) \leq 0$$

Let Z be the set of those elements from $C[-h, 0]$ for which $\dot{V} = 0$ and Q is the greatest invariant set situated in Z. Then all solutions of (5.21) tend to Q for $t \to \infty$. In particular, if the set Q has the only zero element then the trivial solution of (5.21) is asymptotically stable.

Set forth the theorem extending the *Matrosov stability criterion founded on the use of two Liapunov functions* [108(5), 162(10)]. Consider problem (5.1). Let $V(t, \varphi)$ and $W(t, \varphi)$ be continuous functionals defined on $I \times Q_H$ and denote $V(t)$ by $V(t, x_t)$ and $W(t)$ by $W(t, x_t)$. The derivative $\dot{W}$ is called *integrally unbounded* in a set $G \subseteq Q_H$ if for any number $B > 0$ there exist a number $T(B) > 0$ and a continuous function $\xi(t)$ such that uniformly in $x_t = x(t + \theta) \in G$ for $t \geq t_0$

$$\dot{W} \leq \xi(t), \quad \int_t^{t+T(B)} \xi(s)\, ds \leq -B \tag{5.22}$$

Denote $d(\varphi(\theta), G)$ the distance between the element $\varphi(\theta) \in C[-h, 0]$ and the set $G \subseteq C[-h, 0]$.

Theorem 5.7 [108(5), 162(10)]. *It is necessary and sufficient for uniform asymptotic stability of the trivial solution of (5.1) that there exist the functionals $V(t, \varphi)$ and $W(t, \varphi)$ such that*

(1) $\omega_1(|\varphi(0)|) \leq V(t, \varphi) \leq \omega_2(\|\varphi(\theta)\|)$, $t \geq t_0$, $\varphi(\theta) \in Q_H$, $t_0 \in R_1$;
(2) $\dot{V} \leq \bar{\omega}_3(x_t) \leq 0$, where $\bar{\omega}_3(\varphi)$ is a continuous functional defined on Q_H;
(3) $|W(t, \varphi)| \leq L$, $t \geq t_0$, $\varphi(\theta) \in Q_H$, $t_0 \in R_1$;
(4) for any $\mu \in (0, H)$ there exists a $\rho > 0$ such that the derivative $\dot{W}$ is integrally unbounded in the set $E(\mu, \rho) \subseteq C[-h, 0]$, where

$$E(\mu, \rho) = \{\varphi(\theta) \in Q_H,\, d(\varphi(\theta), Q(\bar{\omega}_3 = 0)) \leq \rho,\, \mu \leq \|\varphi(\theta)\| \leq H\}$$

Here $Q(\bar{\omega}_3 = 0) = \{\varphi \in Q_H, \bar{\omega}_3(\varphi) = 0\}$. In addition, the attraction domain of the trivial solution is the sphere S_K, where $K < H$ and $\omega_2(K) \leq \omega_1(H)$.

A general outline of the proof follows. Conditions (1) and (2) ensure the hit of any trajectory $x(t)$ in the neighbourhood $E(\mu, \rho)$ of the set Q. By virtue of (3) and (4) the trajectory necessarily leaves $E(\mu, \rho)$ after a finite time. Then $\dot{V}$ will be negative. So the function V decreases and the trajectory $x(t)$ approaches the origin. For a full proof of Theorem 5.7 see Kolmanovskii and Nosov [108(5)] and Nosov [162(10)].

EXAMPLE 5.2. Consider the equation
$$\dot{x}(t) = -ax(t) + b(t)x(t-h) + c(t)y(t) \\ \dot{y}(t) = -\gamma c(t)x(t), \quad t \geq 0, \quad h \geq 0 \tag{5.23}$$

Show that sufficient conditions of uniform asymptotic stability are $(-a + |b(t)|) \leq c_1 < 0$, $c_3 \geq |c(t)| \geq c_2 > 0$, $\gamma > 0$. Introduce the functional

$$V(t, x_t, y_t) = \gamma x^2(t) + y^2(t) + \gamma a \int_{t-h}^{t} x^2(s)\, ds$$

We have
$$\dot{V} \leq \gamma(-a + |b|)(x^2(t) + x^2(t-h)) \leq 0$$

The set $Q(\bar{\omega}_3 = 0) \subseteq C[-h, 0]$ is, in our case, the set of functions $(\varphi, \psi) \in C[-h, 0]$ such that $\varphi(0) = \varphi(-h) = 0$. The set $Q(\bar{\omega}_3 = 0)$ does not coincide with the origin. As a second functional we take

$$W(t, x_t, y_t) = -y(t)x(t) \operatorname{sgn} c(t) \tag{5.24}$$

This functional is bounded in any sphere Q_H and also
$$\dot{W} = -x(t)(-\gamma c(t)x^2(t)) \operatorname{sgn} c(t) - y(t) \operatorname{sgn} c(t) \\ \cdot [-ax(t) + b(t)x(t-h) + c(t)y(t)]$$

The set $E(\mu, \rho)$ is defined by the inequalities $|x(t)| \leq \rho$, $|x(t-h)| \leq \rho$, $\mu^2 \leq y^2(t) \leq H^2$. In the set $E(\mu, \rho)$ we have

$$\dot{W} = -c_2\mu^2 + a\rho H + |b(t)|\rho H + \gamma c_3 \rho^2 \leq -c_2 \mu^2/2$$

where ρ is such that
$$a\rho H + |b(t)|\rho H + \gamma c_3 \rho^2 \leq 2a\rho H + \gamma c_3 \rho^2 \leq c_3 \mu^2/2.$$

Hence the derivative $\dot{W}$ is integrally unbounded in the set $E(\mu, \rho)$. Asymptotic stability of the trivial solution of (5.23) follows now from Theorem 5.7.

5.6. Global Stability

Let the map $f(t, \varphi)$, $f: R_1 \times C[-h, 0] \to R_n$ be continuous and Lipschitzian in $\varphi \in Q_H$ for arbitrary H. Further, $|f(t, \varphi)| \leq M_H$, $t \in R_1$, $\varphi(\theta) \in Q_H$.

Definition 5.1. *The trivial solution of (5.1) is globally uniformly asymptotically stable if it is uniformly stable and if for any $\gamma > 0$ and a sphere Q_K there* exists an $l(\gamma, K)$ such that $|x(t, t_0, \varphi)| \leq \gamma$ for any $\varphi(\theta) \in Q_K$ and $t \geq t_0 + l(\gamma, K)$.

Theorem 5.8 [108(5)]. *Let there exist two functionals $V(t, \varphi)$ and $W(t, \varphi)$ satisfying the conditions of Theorem 5.7 in any Q_H and*

$$\omega_1(|\varphi(0)|) \leq V(t, \varphi), \qquad \omega_1(s) \to \infty, \quad s \to \infty \qquad (5.25)$$

Then the trivial solution of (5.1) is globally uniformly asymptotically stable.

Notice that under condition (5.25) and other assumptions sufficient for asymptotic stability (see Theorems 5.1, 5.3 and 5.4) the solution is also globally asymptotically stable.

5.7. Survey of Other Results. Exponential Stability

The stability problems for RFDEs are studied in different ways. *Stability in the first approximation* in various senses is studied by Halanay [79(4)], Hale [81(1)] and Kolmanovskii and Nosov [108(5)]. *Stability under steady acting disturbances* is considered by Germaidze and Krasovskii [63]. *Instability* theorems are proved by Shimanov [205(1)]. *Critical cases* (i.e., the cases in which some roots of the characteristic function lie on the imaginary axis) are also studied [167, 178, 205(2)]. The Klimushev papers [101] are devoted to the study of the *stability of singular perturbed RFDEs*. For applications of the vector *Liapunov functional* to the stability of connected systems see Gromova and Markos [73]. Also available is a survey devoted to the Liapunov direct method for RFDEs [3, 108(5)].

Mention some results about exponential stability. The trivial solution of (5.1) is *exponentially stable* if any solution $x(t, t_0, \varphi)$ of (5.1) satisfies the inequality

$$|x(t, t_0, \varphi)| \leq B\|\varphi(\theta)\| \exp(-\alpha(t - t_0))$$
$$B > 0, \quad \alpha > 0, \quad t \geq t_0, \quad \|\varphi(\theta)\| \leq H_1 \leq H \qquad (5.26)$$

A necessary and sufficient condition [114(5)] of exponential stability of the trivial solution of (5.1) is the existence of a functional $V(t, \varphi)$ such that

$$C_1\|\varphi(\theta)\| \leq V(t, \varphi) \leq C_2\|\varphi(\theta)\|$$
$$\dot{V} \leq -C_3\|x_t\|, \qquad |V(t, \varphi) - V(t, \psi)| \leq C_4\|\varphi(\theta) - \psi(\theta)\| \qquad (5.27)$$

where C_i are some positive constants. Note that for a linear RFDE the exponential stability is equivalent to the uniform asymptotic stability. Further, for the linear autonomous RFDE (5.17), asymptotic stability implies uniform asymptotic stability and, hence, exponential stability. It is interesting to notice that for linear autonomous NFDEs, asymptotic stability does not, in general, imply uniform stability. For example [78(1)], the equation $\dot{x}(t) + \dot{x}(t-h) + x(t) = 0, h > 0$, has a solution tending to zero for $t \to \infty$ as t^{-v} where $v > 0$ (see Chapter 1 §5). This means that the trivial solution is not exponentially stable. Hence, asymptotic stability may not be uniform in the initial function.

§6. CONSTRUCTION OF LIAPUNOV-KRASOVSKII FUNCTIONALS

6.1. Linear Autonomous Equations

Krasovskii has considered equations

$$\dot{x}(t) = Cx(t) + Bx(t-h) + \int_{-h}^{0} D(\theta)x(t+\theta)\,d\theta, \qquad x \in R_n \quad (6.1)$$

A quadratic functional is constructed as [114(6), 183(2)]

$$V(x_{t_0}) = x'(0)\alpha x(0) + \int_{-h}^{0} x'(s)\beta(s)x(s)\,ds$$

$$+ \int_{-h}^{0} \int_{-h}^{0} x'(s_1)\gamma(s, s_1)x(s)\,ds\,ds_1 \quad (6.2)$$

Here α is a constant $(n \times n)$ matrix and β, γ are $(n \times n)$ matrices with piecewise differentiable elements. The derivative $\dot{V}$ of functional (6.2) along the paths of Eq. (6.1) equals

$$\dot{V} = x'(0)\omega x(0) + x'(-h)\lambda x(0)$$

$$+ \int_{-h}^{0} x'(s)v(s)x(s)\,ds + \int_{-h}^{0} x'(s)\varphi(s)x(-h)\,ds$$

$$+ \int_{-h}^{0} \int_{-h}^{0} x'(s_1)\varepsilon(s, s_1)s(s)\,ds\,ds_1 \quad (6.3)$$

Theorem 6.1 [114(6)]. *Let the solutions of Eq. (6.1) be asymptotically stable. Then for any functional W of the form (6.3) there exists the unique functional V of the form (6.2) such that $\dot{V} = W$. Besides, if the functional W is negative definite, then the functional V is positive definite. Equations for matrices α, β, γ depending on $\omega, \lambda, v, \varphi$ and ε are derived by Repin* [183(2)].

§6. Construction of Liapunov-Krasovskii Functionals

Quadratic Liapunov functionals giving *necessary and sufficient stability conditions* for the matrix equation

$$\dot{x}(t) = Ax(t) + Bx(t - h), \qquad x \in R_n \tag{6.4}$$

are constructed by Castelan [34, 35]. These functionals are obtained by passage to the limit in Liapunov functions derived in the usual way for difference equations which approximate RFDE (6.4). Notice that this algorithm is complicated enough and the problem of construction of more simple functionals remains actual.

Tzar'kov and Engel'son study the problem of construction of a Liapunov-Krasovskii functional for linear autonomous and periodic systems [231(1), 232(2)]. The functional is constructed in the form

$$V(t, \varphi) = \iint_Q \varphi(s)\varphi(s_1) \, d\mu^t$$

Here μ^t is a symmetric measure defined on the square $Q = [-h, 0] \times [-h, 0]$ for every t. A system of equations is derived for this measure. The equation $\dot{x}(t) = +ax(t - 1)$ is considered as an example by Tzar'kov [231]. Liapunov-Krasovskii functional giving the necessary and sufficient condition of asymptotic stability (i.e., the condition $-\pi/2 < a < 0$, see §3) is

$$V(\varphi) = \varphi^2(0) + a^2 \int_{-1}^0 \varphi^2(s) \, ds + (a^2 + 2) \left[\int_{-1}^0 g(s) \cos as \, ds \right.$$

$$\left. + \frac{\cos a}{1 + \sin a} \left(\int_{-1}^0 g(s) \sin as \, ds - \frac{\varphi^2(0)}{2a} \right) \right]$$

where

$$g(s) = -\varphi(0)\varphi(s) - a \int_{-1}^0 \varphi(\theta)\varphi(-1 - s - \theta) \, d\theta$$

6.2. Stability of a Chemical Reactor Closed by a P Controller

Investigate the stability of the chemical reactor described in Chapter 1, §1. Assume that the controller is a P controller and the actuator has dead zone, saturation and hysteresis loops. A structural scheme of this feedback system is represented in Fig. 2.20. The nonlinear element describing the actuator may be defined in different ways. The output $u(t)$ may be considered as a one-valued functional depending on previous history $x(s)$, $0 \le s \le t$; i.e., $u(t) = F(x(t + \theta))$, $-t \le \theta \le 0$. The distinctive feature of such a description is that

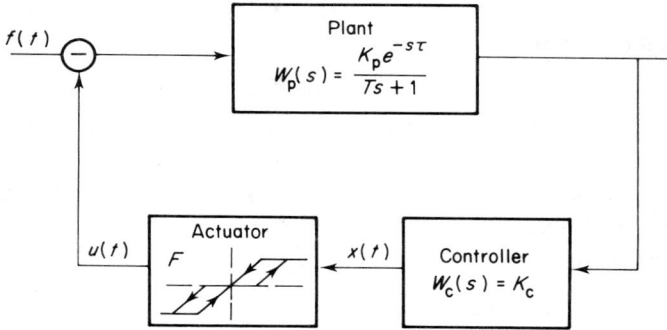

Fig. 2.20. Structural scheme of feedback system for chemical reactor.

the functional has unbounded delay for $t \to \infty$. It is obvious that F satisfies the Lipschitz condition:

$$|F(\varphi(t+\theta)) - F(\psi(t+\theta))| \le r\|\varphi(t+\theta) - \psi(t+\theta)\|_{C[0,t]}$$

or

$$|F(\varphi(t+\theta)) - F(\psi(t+\theta))|^2 \le \int_0^\infty |\varphi(t-s) - \psi(t-s)|^2 \, d_s R(s, \varphi, \psi)$$

$$\int_0^\infty d_s R(s, \varphi, \psi) \le r^2 \tag{6.5}$$

The system represented in Fig. 2.20 is described by the equation $T\dot{x}(t) + x(t) = -K_c K_p F(x(t - \tau + \theta)) + K_c K_p y(t - \tau)$. Let $\bar{y}(t)$ be a fixed input signal, $\bar{x}(t)$ corresponding to $\bar{y}(t)$ be a disturbed solution and $(\bar{x}(t) + z(t))$ be a disturbed solution. The disturbance $z(t)$ satisfies the equation

$$T\dot{z}(t) + z(t) = -K_c K_p F_1(z(t - \tau + \theta)) \tag{6.6}$$

where

$$F_1(z(t - \tau + \theta)) = F(\bar{x}(t - \tau + \theta) + z(t - \tau + \theta)) - F(\bar{x}(t - \tau + \theta))$$

$$F_1^2(z(t - \tau + \theta)) \le \int_0^\infty z^2(t - \tau + \theta) \, d_s R(s, \bar{x}_s, z_s) \tag{6.7}$$

Consider the functional

$$V = Tz^2(t) + K_c K_p r^{-1} \int_0^\infty [d_s R(s, \bar{x}_s, z_s)] \int_{t-\tau+s}^t z^2(s_1) \, ds_1$$

§6. Construction of Liapunov-Krasovskii Functionals

The derivative of this functional is

$$\dot{V} = 2z(t)[-z(t) - K_c K_p F_1(z(t - \tau + \theta))]$$

$$+ K_c K_p r^{-1} z^2(t) \int_0^\infty d_s R(s, \bar{x}_s, z_s) - K_c K_p r^{-1}$$

$$\cdot \int_0^\infty z^2(t - \tau + s) d_s R(s, \bar{x}_s, z_s) \leq -2z^2(t) + K_c K_p r z^2(t)$$

$$+ K_c K_p r^{-1} F_1^2(z(t - \tau + \theta)) + K_c K_p r^{-1} z^2(t) r^2 - K_c K_p r^{-1}$$

$$\cdot \int_0^\infty z^2(t - \tau + s) d_s R(s, \bar{x}_s, z_s) \leq 2z^2(t)[-1 + K_c K_p r]$$

If $K_c K_p r < 1$, then all hypothesis of Theorem 5.4 are fulfilled. Hence, the trivial solution of Eq. (6.6) will be asymptotically stable. In other words, the closed system of a reactor regulation will be stable.

6.3. Scalar Nonlinear Equations

Consider the RFDE

$$\dot{x}(t) = -\int_0^\infty x(t - s) \, dk_0(s) + a(t, x_t), \qquad t \geq 0$$
$$x(\theta) = \varphi(\theta), \qquad -\infty < \theta \leq 0, \qquad \|\varphi\|_B < \infty \tag{6.8}$$

Kernel $k_0(s)$ has bounded variation on $[0, \infty]$. The initial function $\varphi(\theta) \in CB[-\infty, 0]$; i.e., it is continuous and bounded on $(-\infty, 0]$. The continuous functional $a(t, \varphi): [0, \infty) \times CB[-\infty, 0] \to R_1$ is such that $a(t, 0) = 0$ and

$$|a(t, \varphi) - a(t, \psi)|^2 \leq \int_0^\infty |\varphi(-s) - \psi(-s)|^2 \, dR(s)$$
$$\varphi, \psi \in CB[-\infty, 0], \qquad r^2 = \int_0^\infty dR(s) < \infty \tag{6.9}$$

Designate α_{i0} and β_{i0} as in §3

$$\alpha_{i0} = \int_0^\infty s^i |dk_0(s)|, \qquad \beta_{i0} = \int_0^\infty s^i \, dk_0(s)$$

Let the kernel $k_0(s)$ have a jump $b > 0$ in zero and also

$$b > \int_{+0}^\infty |dk_0(s)| + r, \qquad \alpha_{10} + \int_0^\infty s \, dR(s) < \infty \tag{6.10}$$

Then the trivial solution of Eq. (6.8) is asymptotically stable. For the proof, consider the functional

$$V(t, x_t) = x^2(t) + \int_{+0}^{\infty} dk_0(s) \int_{t-s}^{t} x^2(s_1)\, ds_1 + r^{-1} \int_{0}^{\infty} dR(s) \int_{t-s}^{t} x^2(s_1)\, ds_1 \tag{6.11}$$

The derivative of the functional (6.11) is

$$\dot{V} = x^2(t)\left(-2b + \int_{+0}^{\infty} |dk_0(s)| + r\right)$$

$$- \int_{+0}^{\infty} x^2(t-s)|dk_0(s)| - r^{-1} \int_{0}^{\infty} x^2(t-s)\, dR(s)$$

$$+ 2x(t)\left[a(t, x_t) - \int_{+0}^{\infty} x(t-s)\, dk_0(s)\right]$$

$$\leq -2x^2(t)\left[b - \int_{+0}^{\infty} |dk_0(s)| - r\right] \tag{6.12}$$

Hence, by virtue of Theorem 5.4, the trivial solution is asymptotically stable under conditions (6.10).

Investigate the similarly nonlinear equation

$$\dot{x}(t) = -\int_{0}^{\infty} x(t-s)\, d_s k_0(t, s) + a(t, x_t), \qquad t \geq 0$$
$$x_0(\theta) = \varphi(\theta) \tag{6.13}$$

Here $k_0(t, s) = 0$ for $t \geq 0$, $s = 0$ and

$$k_0(t, s) = \int_{0}^{s} f(t, s_1)\, ds_1 + \sum_{i=1}^{\infty} \alpha_i(t) e(s - \tau_i(s))$$

$t \geq 0$, $s > 0$, $\tau_1(t) \equiv 0$, $\alpha_1(t) \neq 0$, $2\dot{e}(s) = 1 + \operatorname{sgn} s$

The function $k_0(t, s)$ has for all t a bounded variation in $s \in [0, \infty)$. Assume that the functions $\alpha_i(t)$ are continuous and that $\tau_i(t)$ are continuously differentiable. Function $f(t, s)$ is continuous and $f(t, s) \equiv 0$ for $t \leq 0$, $s \geq 0$. There exists a constant $L > 0$ and a bounded nondecreasing function $g(s)$ such that for all $t \geq 0$ the series $\sum_{i=1}^{\infty} |\alpha_i(t)|$ converge uniformly and $\sum_{i=1}^{\infty} |\alpha_i(t)| < L$,

$$\int_{s_1}^{s_2} |f(t, s)|\, ds \leq g(s_2) - g(s_1), \qquad \forall s_1, s_2, \ 0 < s_1 < s_2 \, \infty$$

Let the Eq. (6.12) coefficients meet the formulated conditions and also let

§6. Construction of Liapunov-Krasovskii Functionals

(a) $0 \leq \dot{\tau}_i \leq q < 1$ for all i (hence functions $t - \tau_i(t)$ have inverse differentiable function $k_i(t)$).

(b)
$$\beta_1 = \inf\bigg[2\alpha_1(t) - \int_0^\infty |f(t+s,s)|\,ds - \int_0^\infty |f(t,s)|\,ds - \sum_{i=2}^\infty (|\alpha_i(t)|$$
$$+ |\alpha_i(k_i(t))\dot{k}_i(t)|)\bigg] > 0.$$

(c)
$$\sup_{t \geq 0}\bigg[\int_0^\infty ds \int_t^{t+s} |f(t_1,s)|\,dt_1 + \int_0^\infty s\,dR(s) + \sum_{i=2}^\infty \int_t^{k_i(t)} |\alpha_i(s)|\,ds\bigg] < \infty.$$

Then the trivial solution of (6.13) is asymptotically stable. For the proof, consider the functional

$$V = x^2(t) + \int_0^\infty ds \int_t^{t-s} |f(t_1,s)|x^2(t_1 - s)\,dt_1$$
$$+ \sum_{i=2}^\infty \int_t^{k_i(t)} |\alpha_i(s)|x^2(s - \tau_i(s))\,ds + r^{-1}\int_0^\infty dR(s)\int_{t-s}^t x^2(t_1)\,dt_1$$

(6.14)

From (6.14) and (a)-(c) it follows that

$$x^2(t) \leq V \leq C_1\|x_t\|_B^2$$

$$\dot{V} \leq -x^2(t)\bigg[2\alpha_1(t) - \int_0^\infty |f(t+s,s)|\,ds - 2r - \int_0^\infty |f(t,s)|\,ds$$
$$- \sum_{i=2}^\infty |\alpha_i(t)| - \sum_{i=2}^\infty |\alpha_i(k_i(t))\dot{k}_i(t)|\bigg] \leq -\beta_1 x^2(t)$$

So, because of Theorem 5.3, the trivial solution of Eq. (6.13) is asymptotically stable.

For the equation $\dot{x}(t) = -a(t)x(t) + b(t)x(t-h)$, conditions (a)-(c) of asymptotical stability take the form $\inf_t a(t) > \sup_t |b(t)|$, $t \geq 0$. The cases in which equations of the first approximation have no jump at zero are more complicated. They are studied on the basis of degenerated functionals in Chapter 3, §4.

6.4. Nonlinear Equations of Second Order

Some functionals for second-order equations are constructed by Kolmanovskii [105(11)] and Kolmanovskii and Nosov [108(5)]. In view of their awkwardness we study here only the system

$$\dot{x}(t) = y(t), \qquad \dot{y}(t) = -\alpha(t)y(t) + a(t)x(t-h) \qquad (6.15)$$

Assume that $0 < \Delta \leq a(t) \leq C_1$, $0 < \alpha(t) \leq C_2$,

$$2\alpha(t) \geq ha(t) + \int_t^{t+h} a(s)\,ds, \qquad \dot{a}(t) \leq 0 \qquad (6.16)$$

Then the trivial solution of Eq. (6.15) is stable. For the proof, consider the functional

$$V = a(t)x^2(t) + y^2(t) + \int_{t-h}^t dt_1 \int_{t_1}^t [y(t_1) - y(t_2)]^2 a(t_2)\,dt_2$$
$$+ \int_{t-h}^t a(t_1 + h)\,dt_1 \int_{t_1}^t y^2(t_2)\,dt_2$$

It may be verified that

$$C_3(x^2(t) + y^2(t)) \leq V \leq C_4(\|x_t\|^2 + \|y_t\|^2)$$

$$\dot{V} \leq \dot{a}(t)x^2(t) + y^2(t)\left[-2\alpha(t) + ha(t)] + \int_t^{t+h} a(s)\,ds\right]$$

Now stability follows from Theorem 5.3.

If both inequalities (6.16) are strict then by Theorem 5.3, we have asymptotic stability. If only the first inequality (6.16) is strict, then asymptotic stability may be proved by means of the Matrosov criterion 5.7. The set Q ($\bar{\omega}_3 = 0$) in this criterion for the functional V consists of elements $y_t(\theta) \in C[-h, 0]$ such that $y(t) = 0$. Define the second functional

$$W = y(t)x(t-h) \qquad (6.17)$$

Functional (6.17) is bounded in any sphere Q_H and

$$\dot{W} = -\alpha(t)y(t)x(t-h) - a(t)x^2(t-h) - y(t)y(t-h)$$

In the set $E(\mu, \rho)$ (i.e., for $\mu^2 < \|x_t\|^2 \leq H^2$ and $|y(t)| \leq \rho$) we have

$$\dot{W} \leq -\Delta\mu^2 + \rho H \sup_{t \geq 0} |\alpha(t)| + \rho H \leq -\Delta\mu^2/2$$

Here ρ is such that

$$\rho H \sup_{t \geq 0} |\alpha(t)| + \rho H \leq \Delta\mu^2/2$$

Now the asymptotic stability of the trivial solution of Eq. (6.15) follows from Theorem 5.7.

§7. Stability of Nuclear Reactors

In Sinha [210] stability of some *third- and fourth-order equations* is investigated by using Liapunov–Krasovskii functionals. For references devoted to applications of the Liapunov direct method to RFDEs see Alekseevskaja and Gromova [3].

§7. STABILITY OF NUCLEAR REACTORS

7.1. Single-Temperature Reactor with Convective Feedback

A mathematical model of the single-temperature reactor with convective feedback described in §1.1.6 is [70]

$$\dot{N}(t) = [-\alpha T(t - \tau) - \varepsilon(N(t) - N_0)]N(t) \quad (7.1)$$
$$\dot{T}(t) = r[N(t) - N_0 - T(t)]$$

Investigate the stability of the stationary solution $T(t) = 0$, $N(t) = N_0$. The system of the first approximation for (7.1) is

$$\dot{n}(t) = -bn(t) - aT(t - \tau), \qquad \dot{T}(t) = n(t) - T(t) \quad (7.2)$$

where $a = \alpha N_0$, $b = \varepsilon N_0$, $n(t) = N(t) - N_0$, $r = 1$. The characteristic equation for system (7.2) has the form

$$z^2 + (1 - b)z + b + a \exp(-z(\tau)) = 0 \quad (7.3)$$

D subdivision (see Chapter 2, §3.2) for quasi-polynomial (7.3) consists of the straight line $a + b = 0$ and the curve

$$a(\omega) = (1 + \omega^2)\omega/(\omega \cos \omega\tau + \sin \omega\tau)$$
$$b(\omega) = \omega(\omega \sin \omega\tau - \cos \omega\tau)/(\omega \cos \omega\tau + \sin \omega\tau)$$

In Fig. 2.21, D_1 denotes the stability domain for $\tau > 0$ and $D_1 \cup D_2$ is the stability domain for $\tau = 0$. The delay τ has a certain destabilizing effect on system (7.2), and it is necessary to take it into account. According to Theorem 5.5 about stability in the first approximation, system (7.1) is asymptotically stable if its parameters lie in the stability domain D_1 of linear system (7.2) [70].

7.2. Stability of Two Interconnected Reactors

The dynamics of two interconnected reactors exchanging energy by neutrons under the assumption that the internal feedback of each reactor

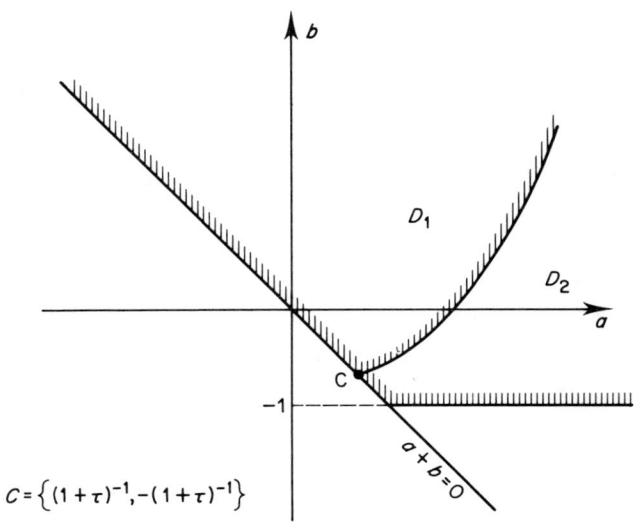

Fig. 2.21. Stability domain for single-temperature nuclear reactor.

may be characterized by negative reactivity factors is described as

$$\dot{N}_1(t) = \frac{\rho_1(t) - \beta_1}{l_1} N_1(t) + \frac{\alpha_{12}}{l_1} N_2(t - \tau_{12}) + \sum_{i=1}^{m_1} \lambda_{i1} C_{i1}$$

$$\dot{N}_2(t) = \frac{\rho_2(t) - \beta_2}{l_2} N_2(t) + \frac{\alpha_{21}}{l_2} N_1(t - \tau_{21}) + \sum_{j=1}^{m_2} \lambda_{j2} C_{j2}$$

$$\dot{C}_{i1}(t) = -\lambda_{i1} C_{i1}(t) + \frac{\beta_{i1}}{l_1} N_1(t), \qquad i = 1, \ldots, m_1$$

$$\dot{C}_{j2}(t) = -\lambda_{j2} C_{j2}(t) + \frac{\beta_{j2}}{l_2} N_2(t), \qquad j = 1, \ldots, m_2$$

$$\rho_1(t) = \rho_{10} - \varepsilon_1 [N_1(t) - N_{10}], \qquad \rho_2(t) = \rho_{20} - \varepsilon_2 [N_2(t) - N_{20}]$$

(7.4)

Here N_i, $\rho_i (i = 1, 2)$ are neutron density and reactivity in the reactor number i, respectively; l_i is neutron lifetime; C_{ik}, λ_{ik} and β_{ik} are concentrations, decay constants and fractions of group number i of retarded neutron emitters in reactor number k, respectively; m_i is a number of groups of retarded neutrons in reactor number i; $-\varepsilon_i$ is a negative power reactivity factor; and $\beta_k = \sum_{i=1}^{m_k} \beta_{ik}$.

Investigate the stability of the stationary solution

$$\begin{aligned} N_k &= N_{k0} > 0, & \rho_k &= \rho_{k0}, & k &= 1, 2 \\ C_{i1} &= C_{i10} > 0, & i &= 1, \ldots, m_1 \\ C_{j2} &= C_{j20} > 0, & j &= 1, \ldots, m_2 \end{aligned}$$

(7.5)

§7. Stability of Nuclear Reactors

Let us introduce the new variables

$$x_k(t) = (N_k(t) - N_{k0})/N_{k0}, \quad k = 1, 2$$
$$y_{i1}(t) = (C_{i1}(t) - C_{i10})/C_{i10}, \quad i = 1, \ldots, m_1$$
$$y_{j2}(t) = (C_{j2}(t) - C_{j20})/C_{j20}, \quad j = 1, \ldots, m_2$$

Equations (7.4) can be written in the form

$$\gamma_1 \dot{x}_1(t) = -b_1 x_1[1 + x_1(t)]$$
$$+ \sum_{i=1}^{m_1} \mu_{i1}(y_{i1}(t) - x_1(t)) - x_1(t) + x_2(t - \tau_{12})$$
$$\gamma_2 \dot{x}_2(t) = -b_2 x(t)[1 + x_2(t)]$$
$$+ \sum_{j=1}^{m_2} \mu_{j2}(y_{j2}(t) - x_2(t)) - x_2(t) + x_1(t - \tau_{21}) \quad (7.6)$$
$$\lambda_{i1}^{-1} \dot{y}_{i1}(t) = x_1(t) - y_{i1}(t), \quad i = 1, \ldots, m_1$$
$$\lambda_{j2}^{-1} \dot{y}_{j2}(t) = x_2(t) - y_{j2}(t), \quad j = 1, \ldots, m_2$$

The parameters of system (7.6) are expressed in an obvious way from the parameters of system (7.4). Consider the Liapunov–Krasovskii functional

$$V = \frac{\gamma_1}{2} x_1^2 + \frac{\gamma_2}{2} x_2^2 + \sum_{i=1}^{m_1} \mu_{i1} \lambda_{i1}^{-1} y_{i1}^2 + \sum_{j=1}^{m_2} \mu_{j2} \lambda_{j2}^{-1} y_{j2}^2$$
$$+ \int_{t-\tau_{12}}^{t} x_1^2(s)\, ds + \int_{t-\tau_{21}}^{t} x_2^2(s)\, ds \quad (7.7)$$

The derivative of functional (7.7) along the trajectories of system (7.6) is

$$\dot{V} = -b_1 x_1^2(t)[1 + x_1(t)] - b_2 x_2^2(t)[1 + x_2(t)]$$
$$- \sum_{i=1}^{m_1} \mu_{i1}[y_{i1}(t) - x_1(t)]^2 - \sum_{j=1}^{m_2} \mu_{j2}[y_{j2}(t) - x_2(t)]^2$$
$$- [x_1(t - \tau_{12}) - x_2(t)]^2/2 - [x_2(t - \tau_{21}) - x_1(t)]^2/2$$

According to physical interpretation, the quantities $N_1(t)$ and $N_2(t)$ are always positive; this implies that $[1 + x_1(t)] > 0$, $[1 + x_2(t)] > 0$. In this subspace all assumptions of Theorem 5.3. are verified. Thus, the stationary solution (7.5) is asymptotically stable for all τ_{12}, τ_{21} and all positive parameters of system (7.4) [70].

§8. MATHEMATICAL MODELS IN IMMUNOLOGY

Problems of mathematical study and simulation of immune processes have been considered by many authors [14(2), 55, 141]. This section deals with some immunological models devoted to the description of virus diseases. These models have been essentially developed and studied by Marchuk [21, 139(1)-139(4)], whose papers we follow.

8.1. Model of Virus Disease

Let us assume that a small quantity of viruses (antigens) has penetrated a human organism and after a certain time reached an organ which they can affect. The viruses begin to multiply and infect the cells and then hit blood and lymphatic nodes. The viruses that have penetrated the lymphatic nodes have some probability of meeting lymphocytes that react to viruses of a given kind. The lymphocyte divides and transforms into an antibody-producing plasmacyte. The formation time of plasmacytes is about 18 to 24 hours. We shall denote the number of viruses in an organism as V, the number of plasmacytes producing antibodies as C, the number of antibodies as F, the relative characteristics of the damaged part of tissue as m, and $\alpha, \beta, \gamma, \mu_c, \mu_f, \rho, \eta, \sigma, \mu_m$ as different positive constants. The variation of the number of viruses in the organism will be described by the equation

$$\dot{V}(t) = \beta V(t) - \mu F(t) V(t) \tag{8.1}$$

The first right-hand term of Eq. (8.1) describes the result of virus division. The second term denotes the number of viruses neutralized by antibodies F. Equations for C, F and m are derived in a similar way. Finally, we arrive at the following system of FDEs describing the dynamics of a virus disease:

$$\begin{aligned}
\dot{V}(t) &= (\beta - \gamma F(t))V(t) \\
\dot{C}(t) &= \xi(m)\alpha F(t-\tau)V(t-\tau) - \mu_c(C - C^*) \\
\dot{F}(t) &= \rho C(t) - \eta\gamma F(t)V(t) - \mu_f F(t) \\
\dot{m}(t) &= \delta V(t) - \mu_m m(t)
\end{aligned} \tag{8.2}$$

Here C^* denotes the constant level of plasmacytes in an organism. The delay τ in the second equation (8.2) represents the time interval from the beginning of stimulation of lymphocytes to the beginning of a mass antibody synthesis. The factor $\xi(m)$ takes into account the fact that antibody production falls if the vitally important organs are seriously damaged. The typical curve of $\xi(m)$ is represented on Fig. 2.22. There is a segment $[0, m^*]$ such that $\xi(m) = 1$ on it; i.e., the functioning of the immunological system does not depend on the

§8. Mathematical Models in Immunology

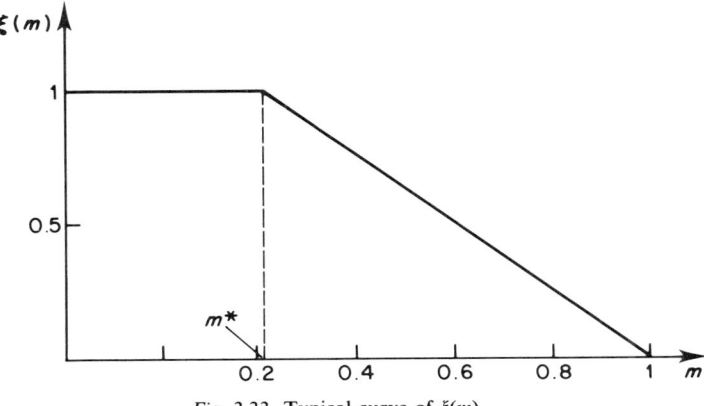

Fig. 2.22. Typical curve of $\xi(m)$.

dynamics of the disease. For a bigger damage (i.e., on the segment $[m^*, 1]$) antibody production falls rapidly. The values of m^* and of the decreasing rates are different for different diseases.

According to the biological interpretation, we consider only the initial data

$$V(t) \equiv 0, \quad t < 0, \; V(0) = V_0, \quad C(0) = C_0$$
$$F(0) = F_0, \quad m(0) = m_0 \tag{8.3}$$

8.2. Analysis of Model (8.2), (8.3)

Theorem 8.1. *Let the initial data (8.3) be positive. Then the solutions of system (8.2) are also positive for all $t \geq 0$.*

From the first equation (8.2) we have

$$V(t) = V_0 \exp\left[\int_0^t (\beta - \gamma F(s)) \, ds\right] \geq 0, \quad t \geq 0 \tag{8.4}$$

On the segment $[0, \tau]$ the second equation (8.2) has a form

$$C(t) = -\mu_c(C - C^*), \; C(0) = C_0.$$

The solution of this equation is $C(t) = C^* + (C_0 - C^*) \exp(-\mu_c t) \geq 0$. Analogously from the third and fourth equations (8.2) we may derive that $F(t) \geq 0$ and $m(t) \geq 0$ on $[0, \tau]$. In the similar way we treat the segments $[\tau, 2\tau]$, etc.

The result of Theorem 8.1 coincides with the biological interpretation of the model. The functions $V(t)$, $C(t)$, $F(t)$ represent the concentrations of different substances and must be positive, and also $0 \leq m(t) \leq 1$.

Theorem 8.1 implies that $C(t) > C^*$ for $t \geq 0$ if $C_0 > C^*$. Really, from the second equation of system (8.2) using the variation-of-constants formula and the inequality $\xi(m(s))F(s - \tau)V(s - \tau) \geq 0$, we have

$$C(t) = C^* + (C_0 - C^*)\exp(-\mu_c t) + \alpha \int_0^t \xi(m(s))F(s - \tau)$$
$$\cdot V(s - \tau)\, ds \geq C^*$$

System (8.2) has two stationary solutions.

Solution 1 ($V_1 = 0$, $C_1 = C^*$, $F_1 = pC^*/\mu_f = F^*$, $m_1 = 0$) describes the state of a *healthy organism*.

Solution 2
$$V_2 = \mu_c(\mu_f \beta - \gamma \rho C^*)/\beta(\alpha \rho - \mu_c \eta \gamma)$$
$$C_2 = (\alpha \mu_f \beta - \eta \gamma^2 \mu_c C^*)/\mu(\alpha \rho - \mu_c \eta \gamma)$$

$F_2 = \beta/\gamma$, $m_2 = \delta V_2/\mu_m$. If these values are positive, they can be interpreted as a state of *chronic disease*.

Theorem 8.2. *Stationary solution 1 is asymptotically stable if*

$$\beta < \gamma F^* \tag{8.5}$$

Proof. If system (8.2) is linearized in the neighbourhood of the stationary solution V_i, C_i, F_i, m_i, $i = 1, 2$, we have

$$\begin{aligned}
\dot{v}(t) &= (\beta - \gamma F_i)v(t) - \gamma V_i f(t) - \gamma f(t)v(t) \\
\dot{s}(t) &= \alpha F_i v(t - \tau) + \alpha V_i f(t - \tau) - \mu_c s(t) + \alpha v(t - \tau)f(t - \tau) \\
\dot{f}(t) &= \rho s(t) - (\eta \gamma V_i + \mu_f)f(t) - \eta \gamma F_i v(t) - \eta \gamma f(t)v(t) \\
\dot{\mu}(t) &= \sigma v(t) - \mu_m \mu(t)
\end{aligned} \tag{8.6}$$

Here we denote $v(t) = V(t) - V_i$, $s(t) = C(t) - C_i$, $f(t) = F(t) - F_i$, $\mu(t) = m(t) - m_i$, $i = 1, 2$. The characteristic equation for the linear part of (8.6) in the neighbourhood of solution 1 can be written in the form

$$D(z) = -(\beta - \gamma F^* - z)(\mu_f + z)(\mu_c + z)(\mu_m + z) = 0$$

Condition (8.5) implies that all zeros of $D(z)$ are negative. Thus, applying Theorem 5.5 on stability in the first approximation to system (8.6) we complete the proof.

Let us investigate now the infection dynamics of a healthy organism. Assume that a small quantity V_0 of viruses has penetrated it at initial moment

§8. Mathematical Models in Immunology

$t = 0$. Consider the component $V(t)$ of the solution of system (8.2) corresponding to the initial data $V_0 > 0$, $C_0 = C^*$, $F_0 = F^*$, $m_0 = 0$.

Theorem 8.3. *Let condition (8.5) hold and also let*

$$0 < V_0 < \text{IB} = \rho\eta^{-1}\beta^{-1}C^* - \mu_f\gamma^{-1}\eta^{-1} \qquad (8.7)$$

Then the component $V(t)$ decreases on the interval $[0, \infty)$.

At $t = 0$ we have $\dot{V}(0) = (\beta - \gamma F^*)V_0 < 0$. Consequently, $V(t)$ decreases on some interval $[0, t_1)$. If $t_1 \neq \infty$, then there is a first moment t_2 such that $\dot{V}(t_2) = 0$ and $\dot{V}(t_2 + \Delta) \geq 0$, $\Delta > 0$. Therefore, it follows from the first equation (8.2) that $\gamma F(t_2) = \beta$, $\gamma F(t_2 + \Delta) \leq \beta$ and $\dot{F}(t_2) \leq 0$. On the other hand, from the third equation (8.2), Theorem 8.1 and Eq. (8.7), we have

$$\dot{F}(t_2) = \rho C(t_2) - \eta\gamma F(t_2)V(t_2) - \mu_f F(t_2) > \rho C^* - \eta\beta V_0 - \mu_f\beta\gamma^{-1} > 0$$

This contradiction proves Theorem 8.3.

The value IB is called an *immunological barrier*. Theorem 8.3 states that if at the initial moment the immunological barrier has not been passed, then the disease is mild and the quantity of viruses penetrating an organism decreases.

Consider now the stationary solution 2. In this case the characteristic equation of the linear part of system (8.6) has the form

$$D(z) = -(\mu_m + z)D_1(z)$$
$$D_1(z) = (-z^3 - az^2 - bz + d + gz\exp(-\tau z) - f\exp(-\tau z)) \qquad (8.8)$$
$$a = \mu_c + \eta\gamma V_2 + \mu_f > 0, \qquad b = \mu_c(\eta\gamma V_2 + \mu_f) - \eta\gamma\beta V_2$$
$$d = \gamma\mu_c\eta\beta V_2 > 0, \qquad g = \alpha\rho V_2, \qquad f = \beta g$$

Theorem 8.4. *If*

$$\mu_c\tau \leq 1, \qquad 0 < (f - d)(a - g\tau)^{-1} < (b - g - f\tau) \qquad (8.9)$$

then stationary solution 2 is asymptotically stable.

Proof. The proof of Theorem 8.4 will follow from Theorem 5.5 about the first approximation stability if we show that Eq. (8.8) has no zeros in the half-plane $\text{Re } z \geq 0$. Applying the Michailov criterion (see Chapter 2, §3.4) to the quasi-polynomial $D_1(z)$, we obtain

$$U(\omega) = \text{Re } D_1(\omega) = a\omega^2 + d - f\cos\omega\tau + g\omega\sin\omega\tau$$
$$V(\omega) = \text{Im } D_1(\omega) = \omega^3 - b\omega + g\omega\cos\omega\tau + f\sin\omega\tau$$

Since $U(0) = d - f < 0$, $V(0) = 0$. The inequalities

$$\beta(a - g\tau) > \gamma\rho C_2(1 - \mu_c\tau) \geq 0, \qquad \mu_c a > b \geq b\tau$$

imply for $\omega > 0$

$$dU/d\omega > \omega[2a - (f\tau + g)\tau - g\tau] > \omega[(a - g\tau) + \tau(b - g - f\tau)] > 0$$

Consequently, there is a value ω_2 such that

$$U(\omega_2) = 0 \quad \text{and} \quad V(\omega_2) < \omega_2[\omega_2^2 - b + g + f\tau] < 0$$

On the interval $(0, \omega_2)$ we have $U(\omega) < 0$ and $V(\omega) < 0$. The vector $D_1(i\omega)$ turns to the angle $\pi/2$ when ω increases from 0 to ω_2. The function $V(\omega) \to \infty$ if $\omega \to \infty$. Thus, there is a value ω_3 such that $V(\omega_3) = 0$ and the vector $D_1(i\omega)$ turns to $\pi/2$ when ω increases from ω_2 to ω_3. When ω increases from ω_3 to ∞, the vector $D_1(i\omega)$ also turns to $\pi/2$ because $V(\omega)$ grows more rapidly than does $U(\omega)$. Finally, the vector $D_1(i\omega)$ turns to $3\pi/2$ when ω increases from 0 to ∞. According to the Michailov criterion, the quasi-polynomial $D_1(z)$ has no zeros in the half-plane $\text{Re } z \geq 0$. Theorem 8.4 is proved.

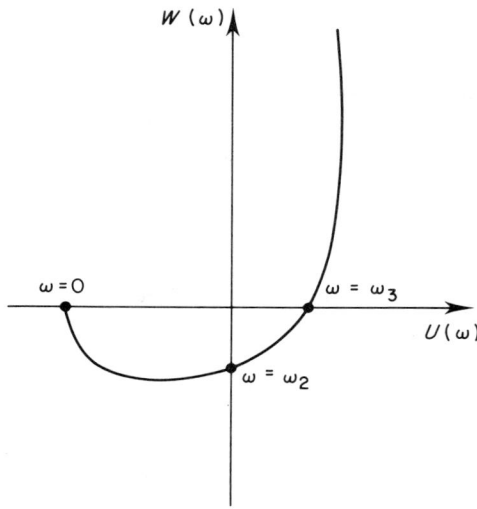

Fig. 2.23. Typical Michailov hodograph for $D_1(i\omega)$.

A typical Michailov hodograph for $D_1(i\omega)$ is shown on Fig. 2.23. Notice that conditions (8.9) depend on the value of delay τ.

8.3. Discussion

Consider now some biological results. On the basis of the previously stated mathematical model we can get typical pictures of virus disease dynamics. If there is a sufficient number of functioning antibodies, the viruses that invade

§8. Mathematical Models in Immunology

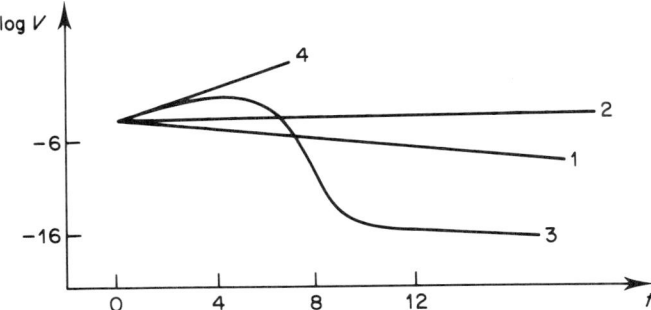

Fig. 2.24. Typical curves of virus disease dynamics: 1, mild case of disease; 2, chronic form of disease; 3, sharp form of disease; 4, mortal form of disease.

an organism will meet with a powerful response and their number will decrease and approach zero. This is a *mild case of a disease* described by Theorems 8.2, 8.3 and shown by curve 1 on Fig. 2.24. In this case the immunological barrier is not passed by the viruses. If the immunological barrier is passed, the number of viruses grows. If, in addition, the number of viruses grows more rapidly than does the number of antibodies neutralizing the viruses, the curve of virus concentration begins to grow exponentially. However, after plasmacytes have formed and mass antibody production has begun the growth of the virus concentration slows down and some time later it falls rapidly. At the same time, there occurs a reproduction of new antibodies whose total number decreases exponentially until the normal immunological level is reached. The damaged part of the organ in which an evolution of the virus population occurred begins to recover exponentially. This is a *sharp form of a disease* (curve 3, Fig. 2.24). If the immunological system is weak and cannot neutralize the viruses, then the number of viruses grows without restriction and the *organism dies* (curve 4, Fig. 2.24).

It may happen, however, that a process of virus multiplication goes on in the organism. The viruses bind with antibodies present in blood plasma. Thus, a balance is established between the number of viruses generated every second and those neutralized by antibodies. Here we deal with a stationary process which can be interpreted as a *chronic or persistent form of a disease* (curve 2, Fig. 2.24).

These four types of virus disease dynamics are illustrated by Table 2.1. The existence of these types of disease dynamics has been confirmed by numerical simulation of model (8.2). Marchuk [139(3)] also considers more complicated models containing 8 or 11 equations, but qualitative analysis of these models is more difficult.

Table 2.1

Disease form	Arise conditions	Peculiarities
1. Mild	$\beta < \gamma F^*$	$V(t) \to 0,\ t \to \infty$
2. Sharp	$\beta > \gamma F^*$	$\dot{V}(t)$ is great
	$\beta - \gamma F^* > 0.33$	
	$\rho g > \eta \gamma f$	
3. Chronic	$0 < \beta - F^*\gamma < 0.33$	$V(t) \to V_2 > 0,\ t \to \infty$
4. Mortal	$\beta > \gamma F^*$	$V(t) \to \infty,\ t \to \infty$
	$g\rho < \eta \gamma f$	

§9. DESIGN OF ADAPTIVE CONTROLLER FOR RETARDED SYSTEMS

Adaptive controllers, and in particular *model references adaptive controllers* (MRAC), are often used for quality control of plants with a wide range of variation of dynamical properties or with considerable noise [108(5), 122]. An MRAC principal scheme is represented in Fig. 2.25. The aim of MRAC design is the construction of an adaptation circuit which provides the stability in phase error $\varepsilon(t) = x(t) - y(t)$. Here $x(t)$ is system output and $y(t)$ is model output.

9.1. Scalar Equations

Let the system be described by the scalar RFDE

$$\dot{x}(t) = -(a + \Delta a + \delta a(t))x(t) + (b + \Delta b + \delta b(t))$$
$$\cdot x(t - h + \Delta h + \delta h(t)) + f(t), \qquad t \geq t_0 \qquad (9.1)$$

Here $\Delta a, \Delta b, \Delta h$ are *a priori* unknown quantities from some domains: $|\Delta a| \leq A_0$, $|\Delta b| \leq B_0$, $|\Delta h| \leq H_0 < h$. Parameters $\delta a(t), \delta b(t), \delta h(t)$ are changed by adaptation circuit and $f(t)$ is an input, $|f(t)| \leq F_0$. Along with RFDE (9.1) consider the reference model

$$\dot{y}(t) = -ay(t) + by(t - h) + f(t), \qquad t \geq t_0 \qquad (9.2)$$

with zero initial data $x(t) = y(t) = \varepsilon(t) = 0$, $t \leq t_0$. For $a > |b|$ the trivial solution of Eq. (9.2) is uniformly asymptotically stable (see §3). At first assume that delays in the system and reference model are equal, i.e., $\Delta h + \delta h(t) = 0$. Then equation for $\varepsilon(t) = x(t) - y(t)$ is

$$\dot{\varepsilon}(t) = -a\varepsilon(t) + b\varepsilon(t - h) + \alpha(t)x(t) + \beta(t)x(t - h)$$
$$t \geq t_0, \quad \alpha(t) = \Delta a + \delta a(t), \quad \beta(t) = \Delta b + \delta b(t) \quad (9.3)$$

§9. Design of Adaptive Controller for Retarded Systems

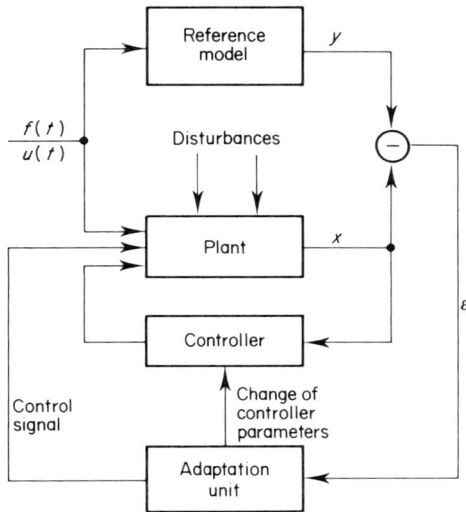

Fig. 2.25. Principal scheme of adaptive controller.

The design of an adaptation circuit may be done on the basis of the Liapunov–Krasovskii functionals method [108(5), 122, 163(2)]. Consider the functional

$$V(t, \varepsilon_t, \alpha, \beta) = \gamma(t)\varepsilon^2(t) + \alpha^2(t) + \beta^2(t)$$

$$+ \left(a\gamma(t) - \frac{1}{2}\dot{\gamma}(t)\right) \int_{t-h}^{t} \varepsilon^2(s)\,ds \qquad (9.4)$$

The derivative of functional (9.4) along the paths of Eq. (9.3) is equal:

$$\dot{V} = (\dot{\gamma} - 2a\gamma)\varepsilon^2(t) + 2\gamma b \varepsilon(t)\varepsilon(t-h)$$

$$+ \left(a\gamma - \frac{1}{2}\dot{\gamma}\right)\varepsilon^2(t) - \left(a\gamma - \frac{1}{2}\dot{\gamma}\right)\varepsilon^2(t-h)$$

$$+ 2\gamma x(t)\varepsilon(t)\alpha(t) + 2\gamma x(t-h)\varepsilon(t)\beta(t)$$

$$+ 2\alpha\dot{\alpha} + 2\beta\dot{\beta} + \left(a\dot{\gamma} - \frac{1}{2}\ddot{\gamma}\right)\int_{t-h}^{t}\varepsilon^2(s)\,ds$$

Take the adaptation circuits of parameters α and β as

$$\dot{\alpha}(t) = (d/dt)(\delta a(t)) = -\gamma(t)x(t)\varepsilon(t), \qquad \alpha(t_0) = \Delta a \qquad (9.5)$$

$$\dot{\beta}(t) = (d/dt)(\delta b(t)) = -\gamma(t)x(t-h)\varepsilon(t), \qquad \beta(t_0) = \Delta b \qquad (9.6)$$

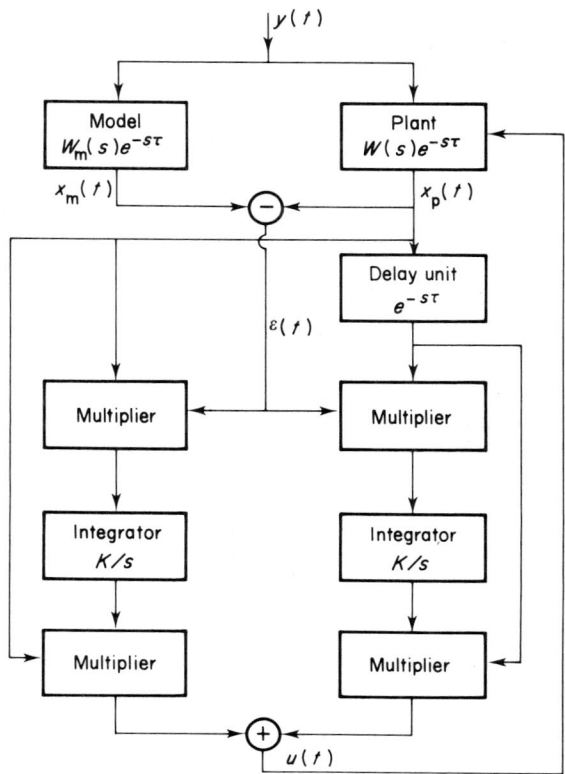

Fig. 2.26. Scheme of adaptive controller for system with delay.

The principal scheme of an adaptive system using algorithm (9.5), (9.6) is represented in Fig. 2.26. The derivative of functional (9.4) along the paths of Eqs. (9.3), (9.5), (9.6) is

$$\dot{V} = (\dot{\gamma} - 2a\gamma)\varepsilon^2(t) + 2\gamma b\varepsilon(t)\varepsilon(t-h)$$

$$+ \left(a\gamma - \frac{1}{2}\dot{\gamma}\right)\varepsilon^2(t) - \left(a\gamma - \frac{1}{2}\dot{\gamma}\right)\varepsilon^2(t-h)$$

$$+ \left(a\dot{\gamma} - \frac{1}{2}\ddot{\gamma}\right)\int_{t-h}^{t} \varepsilon^2(s)\,ds \qquad (9.7)$$

§9. Design of Adaptive Controller for Retarded Systems

Theorem 9.1 [108(5), 163(2)]. *For all $t \geq 0$ let*

$$0 \leq C_1 \leq \gamma(t) \leq C_2, \quad -a + |b| \leq -C < 0$$
$$-C_3 \leq -a\gamma + |b|\gamma + \tfrac{1}{2}\dot{\gamma} \leq -C_4 < 0 \quad (9.8)$$
$$a\dot{\gamma} - \tfrac{1}{2}\ddot{\gamma} \leq 0$$

Then in adaptive system (9.3), (9.5), (9.6) we will have $\varepsilon(t) \to 0$ $t \to \infty$. If, in addition, for any number p, q, $p^2 + q^2 \geq \mu^2 > 0$ and $B > 0$ there exists a number $T(B)$ such that

$$\int_t^{t+T(B)} (py(t) + qy(t-h))^2 \, dt \geq 3B \quad (9.9)$$

then the trivial solution of Eqs. (9.3), (9.5), (9.6) is globally uniformly asymptotically stable and stable under steady acting disturbances.

Proof. According to Theorem 9.1 we get

$$C_5(\varepsilon^2(t) + \alpha^2(t) + \beta^2(t)) \leq V(t, \varepsilon_t, \alpha, \beta)$$
$$\leq C_6 \max_{t-h \leq s \leq t} (\varepsilon^2(s) + \alpha^2(s) + \beta^2(s))$$

Relation (9.7) yields

$$\dot{V} \leq (-a + |b| + \tfrac{1}{2}\dot{\gamma})\gamma(t)\varepsilon^2(t) + (-a\gamma + |b|\gamma + \tfrac{1}{2}\dot{\gamma})\varepsilon^2(t-h)$$
$$\leq -C_4(\varepsilon^2(t) + \varepsilon^2(t-h)) \leq 0$$

Assume now that $\varepsilon(t)$ does not tend to zero. Then the integral from $\dot{V}$ is unbounded but the difference $V(t) - V(0)$ is bounded. Hence $\varepsilon(t) \to 0$, $t \to \infty$. For the proof of the second part of Theorem 9.1 let us use the Matrosov criterion (Theorem 5.7). The equality $\dot{V} = 0$ is possible only for $\varepsilon(t) = \varepsilon(t-h) = 0$. So functional (9.4) satisfies the first two conditions of Theorem 5.7. As a second functional consider

$$W(t, \varepsilon_t, \alpha, \beta) = (\alpha(t)x(t) + \beta(t)x(t-h))\varepsilon(t) \quad (9.10)$$

Denote $E(\mu, \rho)$ the domain

$$E(\mu, \rho) = \{(\varepsilon, \alpha, \beta) \subseteq C[-h, 0] \times R_2, |\varepsilon(t)| \leq \rho,$$
$$|\varepsilon(t-h)| \leq \rho, 0 < \mu^2 < \alpha^2(t) + \beta^2(t) < a_1^2\}, \quad a_1 < a$$

The derivative $\dot{W}$ of the functional W along the paths of Eqs. (9.3), (9.5), (9.6) equals

$$\dot{W} = (\alpha(t)x(t) + \beta(t)x(t-h))[-a\varepsilon(t) + b\varepsilon(t-h) + \alpha(t)x(t)$$
$$+ \beta(t)x(t-h)] + \varepsilon(t)(d/dt)(\alpha(t)x(t) + \beta(t)x(t-h))$$
$$= (\alpha(t)y(t) + \beta(t)y(t-h))^2 + \varepsilon(t)f(\alpha, \beta, x, \dot{x}, \varepsilon(t), \varepsilon(t-h)) \quad (9.11)$$

For $\dot{W}$ in the domain $E(\mu, \rho)$ the estimate holds

$$\int_s^{s+T(B)} \dot{W}(t)\, dt \geq \int_s^{s+T(B)} (py(t) + qy(t-h))^2\, dt - \frac{1}{2}B \geq 2.5B,$$

$$p = \alpha(s), \ q = \beta(s), \qquad p^2 + q^2 \geq \mu^2 > 0$$

Statement of Theorem 9.1 now follows from Theorems 5.7, 5.8.

9.2. Delay Adjustment [162(11)]

Let the reference model be described by (9.2) and let the plant be described by the equation

$$\dot{x}(t) = -ax(t) + bx(t - h + \Delta h + \delta h(t)) + f(t)$$

$$t \geq t_0 \qquad (9.12)$$

If $x(t)$ is a twice continuously differentiable function and $|\Delta + \delta h(t)| < h$ for all $t > t_0$, then $\varepsilon(t) = x(t) - y(t)$ satisfies the RFDE

$$\begin{aligned}\dot{\varepsilon}(t) &= -a\varepsilon(t) + b\varepsilon(t-h) + b[x(t-h+\tau(t)) - x(t-h)] \\ &= -a\varepsilon(t) + b\varepsilon(t-h) + b\tau(t)\dot{x}(t-h) \\ &\quad + b(\tau^2(t)/2)\,\ddot{x}(t-h+\theta\tau(t))\end{aligned} \qquad (9.13)$$

where $\tau(t) = \Delta h + \delta h(t), \ 0 < \theta < 1$.

Consider the following algorithm of delay self-adjustment:

$$\begin{aligned}\dot{\tau}(t) &= (d/dt)(\delta h(t)) \\ &= -b\gamma(t,x)\dot{x}(t-h)\varepsilon(t) - \eta(t,x)[\varepsilon(t)/u(t)]\end{aligned} \qquad (9.14)$$

Here $u(t)$ is a sensitivity function of solution $x(t)$ relative to the following delay variations: $\dot{u}(t) = -au(t) + bu(t-h) + b\dot{x}(t-h), \ u(t) = 0, \ t \geq t_0$. Assume that $|u(t)| > 0$. Then algorithm (9.14) is physically realizable. Algorithm (9.14) is more complicated than (9.5) and (9.16) since, in order for it to be realized, we must know not only $x(t)$ or $x(t-h)$, but also the derivative $\dot{x}(t-h)$. Notice that if a system is described by Eq. (9.1), and the variables $x(t), x(t-h), \dot{x}(t)$ may be measured exactly, then Δa and Δb may be determined algebraically from the equation, and we may then set $\delta a(t) = -\Delta a, \ \delta b(t) = -\Delta b$. Otherwise, under these assumptions the problem of parameter adjustments $\alpha(t)$ and $\beta(t)$ is trivial. But for Eq. (9.12) the knowledge of $x(t), x(t-h), \dot{x}(t)$ and $\dot{x}(t-h)$ is not sufficient in general for the determination of $\tau(t)$ by an elementary method. Therefore, it is meaningful to take into account $\dot{x}(t-h)$ in adaptation algorithm (9.14) for $\tau(t)$.

§9. Design of Adaptive Controller for Retarded Systems

Theorem 9.2 [108(5)]. *Let the first condition of Theorem 9.1 be fulfilled, let the function x(t) be twice continuously differentiable and let*

$$0 \le C_4 \le \eta(t, x) \le C_5, \quad |\ddot{x}(t)| \le C, \quad |u(t)| > 0 \tag{9.15}$$

Then the trivial solution of system (9.13), (9.14) will be uniformly asymptotically stable and stable under steady-acting disturbances for all sufficiently small Δh such that $|\tau(t)| < h$ for $t \ge t_0$.

9.3. Multidimensional Systems

Let the reference model have the form

$$\dot{y}(t) = A(q^0)y(t) + B(q^0)y(t - h) + f(t), \quad y(t) \in R_n \tag{9.16}$$

The system equation is

$$\dot{x}(t) = A(q)x(t) + B(q)x(t - h) + \Phi(t) + f(t), \quad x \in R_n \tag{9.17}$$

Here matrices $A(q)$ and $B(q)$ are linearly dependent on the parameters $q = (q_1, \ldots, q_m)$, and $\Phi(t)$ is the action on the system from the adaptation circuit. We are given

$$q = q^0 + \Delta q, \quad \Delta q = (\Delta q_1, \ldots, \Delta q_m) \tag{9.18}$$

Then Eq. (9.17) has the form

$$\dot{x}(t) = A(q^0)x(t) + B(q^0)x(t - h)$$
$$+ \sum_{i=1}^{m} \left[\frac{\partial A}{\partial q_i} \Delta q_i \, x(t) + \frac{\partial B}{\partial q_i} \Delta q_i \, x(t - h) \right] + \Phi(t) + f(t) \tag{9.19}$$

The problem is to choose a function $\Phi(t)$ such that $\varepsilon(t) \to 0$ for $t \to \infty$. Put

$$\Phi(t) = \sum_{i=1}^{m} \left[\frac{\partial A}{\partial q_i} \delta q_i(t) x(t) + \frac{\partial B}{\partial q_i} \delta q_i(t) x(t - h) \right] \tag{9.20}$$

Equation (9.20) is an adaptation signal for which the functions $\delta q_i(t)$, $i = 1, \ldots, m$, must be determined. For $\varepsilon = x - y$, we get

$$\dot{\varepsilon}(t) = A(q^0)\varepsilon(t) + B(q^0)\varepsilon(t - h)$$
$$+ \sum_{i=1}^{m} \left[\frac{\partial A}{\partial q_i} x(t) + \frac{\partial B}{\partial q_i} x(t - h) \right] \alpha_i(t) \tag{9.21}$$

$$\alpha_i(t) = \Delta q_i + \delta q_i(t)$$

Define functions $\alpha_i(t)$ by the equations

$$\dot{\alpha}_i(t) = \varphi_i(t, \varepsilon_t, x_t) \tag{9.22}$$

Find functions φ_i by using the Liapunov direct method. There exists a quadratic functional

$$V_1(t, \varepsilon(t), \varepsilon_t) = \varepsilon'(t)\Gamma(t)\varepsilon(t) + V_2(\varepsilon_t), \qquad \Gamma' = \Gamma > 0 \qquad (9.23)$$

such that the derivative of functional (9.23) along the paths of the system

$$\dot{\varepsilon}(t) = A(q^0)\varepsilon(t) + B(q^0)\varepsilon(t - h) \qquad (9.24)$$

will be negative-definite $(\dot{V}_1)_{(9.24)} \leq -\omega_3(|\varepsilon(t)|)$, where the subscript indicates that the derivative is taken along the path of Eq. (9.24). The existence of such a functional follows from Krasovskii Theorems 5.3, 6.1. Take now the functional

$$V(t, \varepsilon(t), \varepsilon_t, \alpha(t), \beta(t)) = V_1(t, \varepsilon(t), \varepsilon_t) + \sum_{i=1}^{m} \alpha_i^2(t) \qquad (9.25)$$

The derivative $\dot{V}$ of functional (9.25) along the trajectories of equations (9.21), (9.22) is equal to

$$\dot{V} = (\dot{V}_1)_{(9.24)} + \varepsilon'(t)\Gamma(t)\left[\sum_{i=1}^{m}\left(\frac{\partial A}{\partial q_i}x(t) + \frac{\partial B}{\partial q_i}x(t-h)\right)\alpha_i(t)\right]$$

$$+ \left[\sum_{i=1}^{m}\left(\frac{\partial A}{\partial q_i}x(t) + \frac{\partial B}{\partial q_i}x(t-h)\right)\alpha_i(t)\right]'\Gamma(t)\varepsilon(t)$$

$$+ 2\sum_{i=1}^{m}\alpha_i(t)\varphi_i(t, \varepsilon_t, x_t)$$

Assume now that

$$\varphi_i(t, \varepsilon_t, x_t) = -\varepsilon'(t)\Gamma(t)\left(\frac{\partial A}{\partial q_i}x(t) + \frac{\partial B}{\partial q_i}x(t-h)\right)$$

Then the derivative V will be negative-definite relative to ε_t:

$$\dot{V} = (\dot{V}_1)_{(9.24)} \leq -\omega_3(|\varepsilon(t)|) \leq 0$$

Finally we get the equations of the adaptation circuit

$$\dot{\alpha}_i(t) = -\varepsilon'(t)\Gamma(t)\left[\frac{\partial A}{\partial q_i}x(t) + \frac{\partial B}{\partial q_i}x(t-h)\right]$$

It may be proved that $\varepsilon(t) \to 0$ as $t \to \infty$ (See Kolmanovskii and Nosov [108(5)], §6.1). For the proof of the relation $\alpha(t) \to 0$ some further assumptions must be made as in Theorem 9.1.

§10. STABILITY OF VISCOELASTIC BODIES

This section is devoted to the study of applications of Liapunov direct methods to the investigation of the stability of viscoelastic bodies. General stability statements formulated in terms of the existence of a Liapunov functional could be easily obtained by adaptation of the theorems from §5. But the greatest interest is in constructing new Liapunov functionals and obtaining the effective stability conditions for concrete problems.

10.1. Constitutive Equations of Viscoelastic Bodies

Viscoelastic bodies are characterized by the need to consider their memory effect, i.e., the influence of all past states of bodies from the initial to the current instant. Such phenomena are called hereditary. General features of phenomena under consideration are creep and relaxation. Notice that for a viscoelastic material which has been unstrained at all times prior to t_0, the strain $\varepsilon(t)$ at time t is determined by the stress history $\sigma(\tau)$, $t_0 \leq \tau \leq t$ through the constitutive equation. In accord with Reiner's second axiom of rheology [6], viscoelastic materials are those whose behavior is partly fluidlike and partly solidlike. The principle phenomenon distinguishing viscoelastic substances from purely viscous fluids is the occurrence of strain recovery and viscoelastic substances are distinguished from perfectly elastic solids by the occurrence of creep. It appears that in viscoelastic substances stress produced by strain should relax as in purely viscous fluids, but at a finite rate. In order to obtain the constitutive equation, the *Boltzmann superposition principle* was used. The evolution of stress and strain are governed by the *constitutive equation*

$$\varepsilon(t) = \frac{\sigma(t)}{E} - \int_{t_0}^{t} \sigma(s) K(t, s) \, ds$$

Here $\varepsilon(t)$ is a strain, $\sigma(t)$ a stress, E the Young's modulus and $K(t, s)$ a creep kernel. The *constitutive equation of nonhomogeneous ageing bodies has the form* [6]

$$\varepsilon(t) = \frac{\sigma(t)}{E}(t + \rho(x)) - \int_{t_0}^{t} \sigma(s) K(t + \rho(x), s + \rho(x)) \, ds$$

Here $\rho(x)$ is a function characterizing the heterogeneity of an age. The model of nonhomogeneous ageing bodies generalizes the foregoing models of classical viscoelasticity.

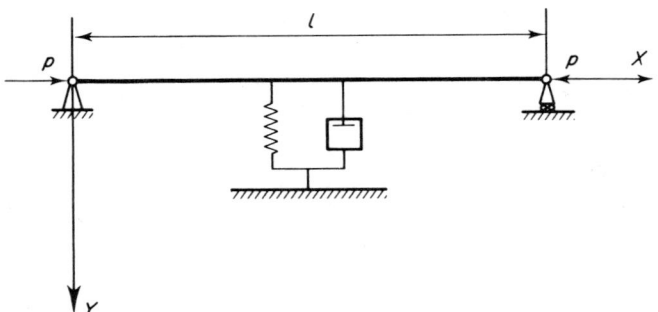

Fig. 2.27. Viscoelastic hinged-hinged bar on viscoelastic foundation.

10.2. Dynamic Stability of a Viscoelastic Bar

Let us derive the stability conditions of a viscoelastic hinged-hinged bar on a viscoelastic foundation (Fig. 2.27). The bar is subjected to an axial load P. Denote $y(t, x)$ as the lateral deflection of the bar from its straight position. The foundation action on the bar is equal to $q = -Cy - C_1\dot{y}$, where $\dot{y} = \partial y(t, x)/\partial t$ and constants $C > 0$, $C_1 \geq 0$. The bar stability is investigated with respect to the initial disturbances of its deflection.

Definition 10.1. *The bar is called stable if for any $\varepsilon > 0$ there exists a $\delta(\varepsilon) > 0$ such that $W(t) = \int_0^l (y(t, x)^2 + \dot{y}(t, x)^2)\, dx < \varepsilon$ as soon as $W(t_0) < \delta(\varepsilon)$.*

The bending moment M is given by

$$M(t, x) = EJ(I - R)y'' = EJ\left(y'' - \int_{t_0}^t r(t - s)y''\, ds\right)$$

$$y'(t, x) = \frac{\partial y(t, x)}{\partial x}$$

Here J is the moment of inertia of the cross section and $r(t - s)$ a relaxation kernel. The *equation of equilibrium* is

$$M'' + Py'' - q = -\mu\ddot{y} \tag{10.1}$$

where μ is the density of the bar material.

Represent the deflection as a Fourier series

$$y(t, x) = \sum_{n=1}^{\infty} a_n(t) \sin nl_1 x, \qquad l_1 = \frac{\pi}{l} \tag{10.2}$$

§10. Stability of Viscoelastic Bodies

Substituting (10.2) in (10.1), we obtain an RFDE for $a_n(t)$

$$\ddot{a}_n + \beta_{1n}\dot{a}_n + \beta_{0n}a_n + \beta_{2n}\int_{t_0}^{t} r(t-s)a_n(s)\,ds = 0$$

$$\beta_{0n} = [C - Pl_1^2 n^2 + EJl_1^4 n^4]\mu^{-1}, \qquad \beta_{1n} = C_1\mu^{-1}$$
$$\beta_{2n} = -\mu^{-1}(nl_1)^4 EJ, \qquad a_n(t_0) = a_{n0}, \quad \dot{a}_n(t_0) = \dot{a}_{n0}. \tag{10.3}$$

Hence the investigation of the stability of the trivial solution of Eq. (10.1) is reduced to the investigation of the stability of Eq. (10.3), studied earlier in this chapter. It is convenient to put $a_n(t) = 0$ for $t < t_0$. Then a solution of (10.3) may be considered as a solution of

$$\ddot{a}_n + \beta_{1n}\dot{a}_n + \beta_{0n}a_n + \beta_{2n}\int_{t_0}^{t} r(s)a_n(t-s)\,ds = 0 \tag{10.4}$$

Consider the Liapunov functionals V_n

$$V_n = \dot{a}_n^2(t) + 2a_n^2(t)\beta_{3n} + \left[\dot{a}_n(t) + \beta_{1n}a_n(t) - \int_{t_0}^{t} r(s)\,ds \int_{t-s}^{t} a_n(t_1)\,dt_1\right]^2$$

$$+ \int_{t_0}^{\infty} r(s)\,ds \int_{t-s}^{t} dt_1 \int_{t_1}^{t} [\dot{a}_n^2(t_2) + \beta_{3n}a_n^2(t_2)]\,dt_2$$

$$\beta_{3n} = \beta_{0n} + \int_{t_0}^{\infty} r(s)\,ds \tag{10.5}$$

The derivatives $\dot{V}_n$ of functionals (10.5) along the trajectories of Eq. (10.4) satisfy the inequality

$$\dot{V}_n \leq -2\left(\beta_{1n} - \int_{t_0}^{\infty} sr(s)\,ds\right)[a_n^2(t)\beta_{3n} + \dot{a}_n^2(t)] \tag{10.6}$$

From the positive-definiteness of functionals (10.5) and the negative-definiteness of functionals (10.6), we arrive at the following *stability conditions*

$$P < \max(EJ\pi^2 l^{-2}, 2\sqrt{EJC}) \qquad C_1\mu^{-1} > \int_{t_0}^{\infty} sr(s)\,ds \tag{10.7}$$

Actually, relations (10.5) and (10.7) yield

$$\sum_{n=1}^{\infty} (a_n^2(t) + \dot{a}_n^2(t)) \to 0, \qquad t \to \infty$$

Hence, by virtue of the Parseval equality, $W(t) \to 0$ for $t \to \infty$. But for any finite time interval, $W(t) = O(W(t_0))$. So the bar is stable under conditions (10.7).

§11. ABSOLUTE STABILITY OF SYSTEMS WITH DELAY

Absolute (in the Lurie–Aizerman sense) stability has been investigated by Popov and Halanay [176]. Razvan's work [181] and the appendix to the Russian translation of it by Lihtarnikov and Jakubovich [130] contain the most important results in this domain. We limit our presentation to the statement of the problem and the formulation of two theorems.

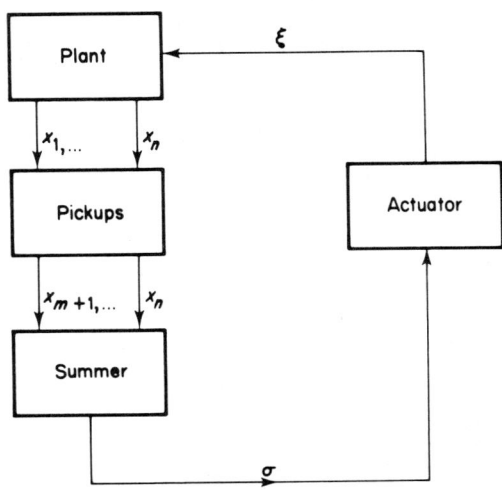

Fig. 2.28. Feedback system with nonlinear actuator.

Consider a feedback system with delay shown in Fig. 2.28. System phase variables are measured by pickups. The adder generates a control signal $\sigma = c_1 x_1 + \cdots + c_n x_n = c'x$, which governs an actuator. The actuator output ξ is equal to $f(\sigma)$ in the case of direct automatic control without delay, i.e., $\xi(t) = f(\sigma(t))$. When there is a delay in the actuator, we assume that $\xi(t) = f(\sigma(t-h))$.

Suppose that plant dynamics is described by the RFDE

$$\dot{x}(t) = Ax(t) + Bx(t-h) + b\xi(t), \qquad x(t) \in R_n$$

Here the matrices A, B and vector b are constant. The system closed by an adder and an actuator with delay is described by the equation

$$\dot{x}(t) = Ax(t) + Bx(t-h) + bf[\sigma(t-h)], \qquad \sigma = c'x \qquad (11.1)$$

§11: Absolute Stability of Systems with Delay

If the actuator has no delay, then we get

$$\dot{x}(t) = Ax(t) + Bx(t-h) + bf(\sigma(t)), \qquad \sigma = c'x \qquad (11.2)$$

As a rule, the characteristic $f(\sigma)$ is nonlinear and its exact determination is difficult. Assume that a continuous characteristic $f(\sigma)$ is defined for all $\sigma f(0) = 0$ and also satisfies the following conditions:

(a) There exist constants $k_1, k_2 > 0$ such that $k_1 \sigma^2 \leq \sigma f(\sigma) \leq k_2 \sigma^2$.
(b) $\int_0^\infty f(\sigma)\, d\sigma = \infty$.

Such characteristics are called *admissible*.

Definition 11.1. *Control system* (11.1) [*or* (11.2)] *is said to be absolutely stable in the angle* $[k_1, k_2]$, *if the trivial solution of Equation* (11.1) [*or* (11.2)] *is globally asymptotically stable for all admissible characteristics* $f(\sigma)$.

For determination of absolute stability conditions of concrete systems one uses the ordinary frequency domain criteria of Popov–Halanay [176]. We now state these criteria.

Theorem 11.1. *Let all the roots of characteristic equation*

$$\det[A + B\exp(-zh) - zI] = 0 \qquad (11.3)$$

lie in the half-plane $\mathrm{Re}\, z \leq -\alpha < 0$. *Also let there exist a number* $q > 0$ *such that*

$$(1/K) + \mathrm{Re}(1 + j\omega q)\exp(-j\omega h)c'(A + B\exp(-j\omega h) - j\omega I)^{-1}$$
$$\cdot b \geq 0, \quad k_1 < k_2 < K, \quad j^2 = -1$$

Then system (11.1) *is absolutely stable in the angle* $[k_1, k_2]$.

Theorem 11.2. *Let all the roots of characteristic Eq.* (11.3) *lie in the half-plane* $\mathrm{Re}\, z \leq -\alpha < 0$ *and also let there exist a number* $q > 0$ *such that*

$$(1/K) + R(1 + j\omega q)c'(A + B\exp(-j\omega h) - j\omega I)^{-1}b \geq 0$$

The system (11.2) *is absolutely stable in the angle* $[k_1, k_2]$.

§12. STABILITY OF LINEAR PERIODIC EQUATIONS

12.1. General Stability Theorem

Consider the equation

$$\dot{x}(t) = L(t, x_t), \qquad t \geq t_0, \quad x_{t_0} = \varphi,$$
$$x_t = x(t + \theta), \qquad -h \leq \theta \leq 0 \tag{12.1}$$

Here $L(t, \varphi)$, $L: R_1 \times C[-h, 0] \to R_n$ is a continuous map linear in φ and ω periodic in t:

$$L(t + \omega, \varphi) = L(t, \varphi)$$
$$L(t, C_1\varphi + C_2\psi) = C_1 L(t, \varphi) + C_2 L(t, \psi), \qquad t \in R_1, \; \varphi, \psi \in C[-h, 0]$$

Basic results about the stability Eq. (12.1) were obtained by Halanay [79(2), 79(4)], Shimanov [205(3)], Stokes [223(1)], Hale [81(4)] and Zverkin [253(3)]. See also El'sgol'tz and Norkin [59], Halanay [79(4)], Kolmanovskii and Nosov [108(5)].

Introduce the shift operator $U(t)$ of period ω on the solution of Eq. (12.1)

$$U(t)\varphi = x_{t+\omega}, \qquad U(t): C[-h, 0] \to C[-h, 0]$$

The operator $U(t)$ maps any function $\varphi(\theta) \in C[-h, 0]$ into the part $x_{t+\omega}$ of the solution of Eq. (12.1) satisfying the initial condition $x_t = \varphi$. The shift operator $U(t)$ is called also a *monodromy operator* [223]. It may be proved [79(2), 223] that for $\omega > h$ and any t the monodromy operator $U(t)$ is *completely continuous* and its spectrum is at most countable and does not depend on t: $\sigma(U(s)) = \sigma(U(t))$ for each s and t. The eigenvalues of operator $U(t)$ also called *characteristic (or Floquet) multipliers of* (12.1).

Theorem 12.1 [223]. *For asymptotic stability of the solutions of Eq. (12.1) it is necessary and sufficient that the spectrum $\sigma(U(t))$ of the monodromy operator $U(t)$ lies inside the unit circle. If all Floquet multipliers ρ_j of Eq. (12.1) are such that $|\rho_j| \leq 1$ and the multipliers ρ_k with $|\rho_k| = 1$ have simple elementary divisors, then the trivial solution of Eq. (12.1) is stable.*

12.2. Particular Classes of RFDEs

Theorem 12.1 reduces the problem of the stability of RFDE (12.1) to the investigation of spectrum $\sigma(U(t))$. Usually the problem of determination or estimation of $\sigma(U(t))$ is very complicated and now there are no significant results in it. But for some particular classes of Eq. (12.1) this problem is

§12. Stability of Linear Periodic Equations

studied better. Consider a *scalar equation with delays divisible by period* [253(3), 253(5)]:

$$\dot{x}(t) + b_0(t)x(t) + b_1(t)x(t - m_1\omega) + b_2(t)x(t - m_2\omega) = 0$$
$$b_i(t + \omega) = b_i(t), \quad i = 0, 1, 2, \quad x(t) \in R_1 \quad (12.2)$$

where m_1 and m_2 are positive integer numbers. Let us look for a solution of the form

$$x(t) = p(t)\exp(\lambda t), \quad p(t + \omega) = p(t) \quad (12.3)$$

Such solutions are usually called *Floquet solutions*. For $p(t)$ we get the equation

$$\dot{p}(t) + g(t, \lambda)p(t) = 0$$
$$g(t, \lambda) = b_0(t) + b_1(t)\exp(-m_1\lambda\omega) + b_2(t)\exp(-m_2\lambda\omega)$$

Integrating this equation we conclude that the function

$$p(t) = C\exp\left\{-\lambda t - \int_0^t g(s, \lambda)\,ds\right\}$$

must be ω periodic. Hence we have the following equation for λ:

$$\lambda\omega + \int_0^\omega g(s, \lambda)\,ds = 0 \quad (12.4)$$

Equation (12.4) is a quasi-polynomial. We know (see Chapter 1, §5) that Eq. (12.4) has at most a countable number of roots and each of them has finite multiplicity. The Floquet solution (12.3) corresponds to every zero of quasi-polynomial (12.4). By Theorem 12.1 for asymptotic stability of Eq. (12.2), it is necessary and sufficient that all the roots λ_j of Eq. (12.4) satisfy the condition Re $\lambda_j < 0$. For systems of equations with delays divisible by period, it is also possible to write out a characteristic equation, but its coefficients depend on the elements of the fundamental matrix of an ordinary differential equation. In this case, simple assertions are not derived [253(5)].

Nevertheless there are some results for equations with quasi-constant coefficients [115]. Thus for a system

$$\dot{x}(t) = (A + \mu g(t, \mu))x(t) + (B + \mu f(t, \mu))x(t - h)$$
$$h > 0, \quad x \in R_n, \quad g(t + \omega, \mu) = g(t, \mu), \quad f(t + \omega, \mu) = f(t, \mu) \quad (12.5)$$

it is shown that for a root λ of the equation

$$\det[A + B\exp(-\lambda h) - I] = 0 \quad (12.6)$$

the particular solution has the form

$$x(t, \mu) = \exp[\alpha(\mu)t]p(t, \mu), \quad \alpha(0) = \lambda, \quad p(t, \mu) = p(t + \omega, \mu)$$

Consequently, if all the roots λ_j of Eq. (12.6) satisfy the condition Re $\lambda_j < 0$, then system (12.5) is asymptotically stable for sufficiently small μ.

Some stability criteria for periodic equations with small parameters are obtained by a Laplace transform. For example, in Valeev [233] it is proved that the trivial solution of the scalar equation $\dot{x}(t) = 2\mu(\cos t)x(t - h)$, $\mu > 0$, is stable if $\sin h > 0$ and unstable if $\sin h < 0$.

12.3. Floquet Theory

Floquet solution of Eq. (12.1) is a solution $x(t)$ such that

$$x_{t+\omega} = \rho x_t, \quad x_t \neq 0, \quad \rho \neq 0, \quad -\infty < t < \infty$$

The number ρ called a Floquet multiplier is also an eigenvalue of the monodromy operator $U(t)$. The *Floquet exponent* λ equals $\lambda = \omega^{-1} \ln \rho$. If ρ is a simple eigenvalue, then the Floquet solution $\psi(t) = p(t) \exp(\lambda t)$, $p(t + \omega) = p(t)$ corresponds to ρ.

Let ρ be an r-multiple eigenvalue of the operator $U(t)$. Then a finite-dimensional subspace $E_r(t) \subset C[-h, 0]$ corresponding to ρ is invariant relative to $U(t)$. The action of the operator $U(t)$ in subspace $E_r(t)$ is described by a matrix. This matrix has Jordan form for appropriate basis. The following r Floquet solutions correspond to an r-multiple eigenvalue

$$\psi_{j1}(t) = p_{j1}(t) \exp(\lambda t)$$

$$\psi_{jm(j)}(t) = [\{1/[m(j) - 1]!\} \, t^{m(j)-1} p_{j1}(t) + \cdots + p_{jm(j)}(t)]e^{\lambda t} \quad (12.7)$$

$$j = 1, \ldots, q, \quad m(1) + \cdots + m(q) = r$$

Here q is the number of Jordan cells of the matrix describing the action of operator $U(t)$ in an invariant subspace $E_r(t) \subset C[-h, 0]$ and $m(j)$ are dimensions of these cells. All the functions $p_{j1}(t), \ldots, p_{jm(j)}(t)$ in formulae (12.7) are ω periodic.

Theorem 12.2. *Any solution $x(t)$ of Eq. (12.1) may be expanded in an asymptotic series over Floquet solutions (12.7) for $t \to \infty$. In other words, for every solution $x(t)$ and a number $\alpha > 0$ there exists the finite sum $\bar{x}(t)$ of Floquet solutions (12.7):*

$$\bar{x}(t) = \sum_{|\rho_i| \geq \exp(-\alpha\omega)} \sum_{j=1}^{q_i} \sum_{l=1}^{m_i} C_{ijl} \psi_{ijl}(t) \qquad (12.8)$$

§12. Stability of Linear Periodic Equations

such that

$$|x(t) - \bar{x}(t)| \leq C \exp(-\beta t), \qquad \beta > \alpha, \quad t \to \infty$$

Coefficients of the expansion (12.8) are determined by the aid of a so-called conjugate system [253(5)]. As an addition to Theorem 12.2 it may be proved that $\bar{x}(t)$ is a solution of some ordinary differential equation. This equation may be written out if Floquet multipliers are known. Hence, solutions of periodic RFDEs are approximated by solutions of ordinary differential equations with precision to functions vanishing faster than any exponential ones. It is possible to develop the perturbation theory of RFDEs similar to those for ordinary differential equations. Notice that in general Theorem 12.2 may not be made more precise; i.e., asymptotic convergence of a series over Floquet solutions cannot be changed by convergence in the usual sense. Hence, a system of Floquet solutions may be *incomplete*.

EXAMPLE 12.1 [253(5)]. Consider a scalar equation

$$\dot{x}(t) = 2\pi(\cos(2\pi t))x(t - 1) \qquad (12.9)$$

Since the mean value of the coefficients equals zero, Eq. (12.4) takes the form $\lambda = 0$. The Floquet solution $\psi_1(t) = \exp(\sin 2\pi t)$ corresponds to the unique root $\lambda_1 = 0$ of Eq. (12.4). Equation (12.9) has no other Floquet solutions, and so the system of Floquet solutions is not complete. Nevertheless, Theorem 12.2 is valid for Eq. (12.9). For any solution $x(t)$ of this equation and arbitrary $\beta > 0$, we have $x(t) = C_1\psi_1(t) + 0[\exp(-\beta t)]$. Consider, for example, the solution x_1 of Eq. (12.9) defined by the initial condition $x_1(t) = 1, 0 \leq t \leq 1$. By the step method (see Chapter 1, §3) we get

$$x_1(t) = \sum_{j=0}^{[t]} \frac{1}{j!} (\sin 2\pi t)^j, \qquad t \geq 0$$

Expanding the solution $\psi_1(t)$ in a Taylor series, we obtain

$$x_1(t) - \psi_1(t) = -\frac{1}{(t+1)!} [\sin(2\pi t)]^{[t+1]} \exp(2\zeta\pi t) = 0\left(\frac{1}{\Gamma(t)}\right)$$

$$0 < \zeta < 1$$

Here $\Gamma(t)$ is gamma function growing quicker than any exponent. Therefore, $\psi_1(t)$ is an asymptotic representation for the solution $x_1(t)$. Notice that $x_1(t) - \psi_1(t)$ is a solution of Eq. (12.9) decreasing more quickly than any exponent. Equations with constant coefficients have no solutions that decrease so quickly. Hence, an RFDE with periodic coefficients may not be reduced to stationary equation by change of variables, unlike ordinary differential equations.

It is interesting to define the classes of equations for which the system of Floquet solutions is complete (or incomplete). Equations with delays divisible by period are being investigated in detail [116, 253(5)]. Consider, for example, the scalar equation

$$\dot{x}(t) = \sum_{j=0}^{m} a_1(t) x(t - j\omega), \qquad a_j(t + \omega) = a_j(t) \qquad (12.10)$$

A system of Floquet solutions of this equation is complete in $C[0, m\omega]$ if the function $a_m(t)$ is of constant sign and does not equal zero in entire segments. For the equation $\dot{x}(t) = a(t)x(t - \omega)$ with alternating signs of $a(t)$, the Floquet system is incomplete.

Chapter 3

Stability of Neutral Functional Differential Equations

§1. STABIILITY OF LINEAR AUTONOMOUS NFDEs

1.1. General Stability Theorems

Consider an NFDE with arbitrary aftereffect $(x(t) \in R_n)$

$$\dot{x}(t) = \int_0^\infty [dK_0(s)]x(t-s) + \int_0^\infty [dK_1(s)\dot{x}(t-s), \qquad t \geq 0 \quad (1.1)$$

with initial data

$$x_0(\theta) = \varphi(\theta), \qquad \dot{x}_0(\theta) = \dot{\varphi}(\theta), \qquad \theta \leq 0 \quad (1.2)$$

Assume that all elements K_l^{ij} of the $(n \times n)$ matrices K_0 and K_1 are functions with bounded variation on $[0, \infty)$ and also

$$\int_0^\infty s\|dK_l(s)\| < \infty, \qquad \sum_{i,j=1}^n \int_0^\infty \|dK_1(s)\| = 1 - \gamma, \qquad \gamma > 0$$

$$\|K_l\| = \sum_{i,j=1}^n |K_l^{ij}| \quad (1.3)$$

The initial function $\varphi(\theta)$ is continuously differentiable and

$$\|\varphi\|_1 = |\varphi(0)| + |\dot{\varphi}(0)| + \left(\int_0^\infty |F_0(t)|^2 \, dt\right)^{1/2} + \left(\int_0^\infty |F_1(t)|^2 \, dt\right)^{1/2} < \infty \quad (1.4)$$

$$F_j(t) = \int_t^\infty [dK_j(s)]\varphi^{(j)}(t-s), \qquad j = 0, 1$$

Under these assumptions problem (1.1), (1.2) has the unique solution $x(t, \varphi)$ such that

$$|x(t, \varphi)| + |\dot{x}(t, \varphi)| \leq C_1\|\varphi\|_1 \exp(C_2 t)$$

where $C_1, C_2 > 0$ are some constants. Hence there exists a Laplace transform $\bar{x}(z)$ of the solution $x(t, \varphi)$. Using relations (2.2.7)–(2.2.10), we get

$$[Iz - \bar{K}_1(z)z - \bar{K}_0(z)]\bar{x}(z) = (I - \bar{K}_1(z))\varphi(0) + \bar{F}_0(z) + \bar{F}_1(z)$$

$$\bar{K}_j(z) = \int_0^\infty e^{-zs}\, dK_j(s), \qquad \bar{F}_j(z) = \int_0^\infty e^{-zt} F_j(t)\, dt, \tag{1.5}$$

$$\bar{x}(z) = \int_0^\infty e^{-zs} x(t, \varphi)\, dt$$

The function

$$\Delta(z) = \det[Iz - \bar{K}_1(z)z - \bar{K}_0(z)] = z^n[1 - \det \bar{K}_1(z)] + 0(|z|^{n-1}) \tag{1.6}$$

is called a characteristic function corresponding to Eq. (1.1). The L_2-stability and asymptotic-stability definitions for Eq. (1.1) coincide with Definitions 2.1.3 and 2.1.4 (Chapter 2, Definitions 1.3 and 1.4) with $\|\varphi\|_1$ instead of ρ.

Theorem 1.1. *Let function (1.6) have no zeros in the half-plane* Re $z \geq 0$ *and kernels $K_j(s)$ satisfy requirements (1.3). Then the trivial solution of Eq. (1.1) is L_2 stable and asymptotically stable.*

Proof. The proof of Theorem 1.1 is similar to the proofs of Theorems 2.2.1 and 2.2.2. On the basis of conditions (1.3) we obtain for Re $z \geq 0$

$$\|\bar{K}_1(z)\| = \left\| \int_0^\infty \exp(-zs)\, dK_1(s) \right\| \leq 1 - \gamma \tag{1.7}$$

From (1.7) follows the existence of a bounded matrix $[I - \bar{K}_1]^{-1}$ such that $\|(I - \bar{K}_1(z))^{-1}\| \leq \gamma^{-1}$, Re $z \geq 0$. Multiplying Eq. (1.5) on $(I - \bar{K}_1(z))^{-1}$, we have

$$[Iz - (I - \bar{K}_1(z))^{-1}\bar{K}_0(z)\bar{x}(z) = \varphi(0) + (I - \bar{K}_1(z))^{-1}(\bar{F}_0(z) + (\bar{F}_1(z)) \tag{1.8}$$

The matrix $\bar{K}_2(z) = (I - \bar{K}_1(z))^{-1}\bar{K}_0(z)$ and the function

$$\bar{F}_2(z) = (I - \bar{K}_1(z))^{-1}(\bar{F}_0(z) + \bar{F}_1(z))$$

are bounded for Re $z \geq 0$. Hence Eq. (1.8) takes the form

$$(Iz - \bar{K}_2(z))\bar{x}(z) = \varphi(0) + \bar{F}_2(z) \tag{1.9}$$

Equation (1.9) is just similar to Eq. (2.2.9). So by virtue of the Plancherel theorem and boundedness of the matrix $(I - \bar{K}_1(z))^{-1}$ we have

$$\int_{-\infty}^{\infty} |\bar{F}_2(i\beta)|^2\, d\beta \leq 2\pi\gamma^{-2}$$

§1. Stability of Linear Autonomous NFDEs

As in Theorem 2.2.1 we can prove that

$$\int_{-\infty}^{\infty} |\bar{x}(i\beta)|^2 \, d\beta < \infty, \qquad \|\varphi\|_1 < \infty$$

Hence, L_2 stability is proved. The asymptotic stability proof is just the same as in Theorem 2.2.2. Theorem 1.1 is proved.

For some NFDEs requirements (1.3) may be weakened. Consider the NFDE

$$\dot{x}(t) = \int_0^\infty [dK_0(s)]x(t-s) + \sum_{m=1}^{\infty} K_m \dot{x}(t-h_m)$$

$$+ \int_0^\infty \lambda(s)\dot{x}(t-s) \, ds, \quad t \geq 0, \quad x \in R_n \quad (1.10)$$

Here K_m are $(n \times n)$ matrices, and the $(n \times n)$ matrix $\lambda(s)$ has absolute integrable elements and

$$\sum_{m=1}^{\infty} \|K_m\| < 1, \qquad \int_0^\infty \|\lambda(s)\| \, ds < \infty \quad (1.11)$$

Conditions (1.11) are weaker than (1.3). The Laplace transform $\bar{x}(z)$ of Eq. (1.10) solution satisfies the relation

$$[\bar{B}(z)z - \bar{K}_0(z)]\bar{x}(z) = \bar{B}(z)\varphi(0) + \bar{F}_0(z) + \bar{F}_1(z),$$

$$B(z) = I - \sum_{m=1}^{\infty} K_m \exp(-h_m z) - \int_0^\infty \lambda(s) \exp(zs) \, ds \quad (1.12)$$

Theorem 1.2. *Let the characteristic function*

$$\Delta(z) = \det[\bar{B}(z)z - \bar{K}_0(z)] \quad (1.13)$$

have no zeros in the half-plane $\operatorname{Re} z \geq 0$ *and kernels* K_0, K_m *and* $\lambda(s)$ *satisfy conditions* (1.3), (1.11). *Then the trivial solution of Eq.* (1.10) *is L_2 stable and asymptotically stable.*

Proof. The function $\bar{x}(z)$ is analytical in $\operatorname{Re} z > 0$ and continuous in $\operatorname{Re} z \geq 0$, consequently,

$$x(t) = \frac{1}{2\pi} \int_{-\infty}^{\infty} \bar{x}(i\beta)e^{i\beta t} \, d\beta$$

By virtue of (1.12)

$$\bar{x}(i\beta) = [\bar{B}(i\beta) - \bar{K}_0(i\beta)]^{-1}[\bar{B}(i\beta)x(0) + \bar{F}_0(i\beta) + \bar{F}_1(i\beta)]$$

Now prove the estimate similar to (2.2.13). Using (1.2) and (1.11), we have

$$\|\bar{B}(i\beta)\| < \infty, \qquad \|\bar{K}_0(i\beta)\| < \infty \qquad (1.14)$$

Therefore

$$\|\bar{B}(i\beta)i\beta - \bar{K}_0(i\beta))^{-1}\| = O(|\beta|^{n-1})/|\Delta(i\beta)| \qquad (1.15)$$

Evaluate $\Delta(i\beta)$. By the assumption made, $\Delta(i\beta) \neq 0$ for all β. Further, take

$$\int_0^\infty \exp((i\beta s)\|\lambda(s)\|\,ds \to 0, \qquad |\beta| \to \infty$$

as the Fourier coefficient of an absolutely integrable function. Then there exists a β_0 such that $\|I - \bar{B}(i\beta)\| \leq \gamma_1 < 1, |\beta| \geq \beta_0$. For $|\beta| \geq \beta_0$ the matrix $\bar{B}(i\beta)$ is nonsingular and also $\|\bar{B}^{-1}(i\beta)\| \leq \gamma_1^{-1}$. But then $|\det \bar{B}(i\beta)| \geq C > 0$, $|\beta| > \beta_0$. So for some $\beta_1 > \beta_0$ we obtain

$$|\Delta(i\beta)| = |\beta|^n |\det \bar{B}(i\beta)| + O(|\beta|^{n-1}) \geq C|\beta|^n/2$$

From here and (1.14), (1.15) it follows that

$$\|(\bar{B}(i\beta)i\beta - \bar{K}_0(i\beta))^{-1}\| \leq C_2 |\beta|^{-1}, \qquad |\beta| \to \infty$$

$$\|\bar{B}(i\beta)i\beta - \bar{K}_0(i\beta))^{-1}\| \leq C_3, \qquad -\infty < \beta < \infty$$

Now for the end of the proof it is sufficient to repeat the proofs of Theorems 2.2.1 and 2.2.2.

REMARK 1.1. Let condition (1.3) be invalid. Then the trivial solution may be unstable even if all the zeros of function (1.6) lie to the left of the imaginary axis. For example, the trivial solution of the equation

$$L(L(L(x))) = 0, \qquad L(x) = \dot{x}(t) - \dot{x}(t-1) + \pi^2 x(t) \qquad (1.16)$$

is unstable in the Liapunov sense. But at the same time all the zeros of the quasi-polynomial corresponding to Eq. (1.16) lie to the left of the imaginary axis [216].

REMARK 1.2. Let requirement (1.3) be fulfilled for some $\gamma > 0$ and let

$$\int_0^\infty \exp(\gamma s)\|dK_j(s)\| < \infty, \qquad j = 0, 1 \qquad (1.17)$$

Further, assume that in the half-plane Re $z \geq -\gamma$, function (1.6) has no zeros. Then, similar to Theorem 2.2.3, it may be proved that $|x(t)| \leq C_\varphi \exp(-\gamma t)$.

1.2. Methods of Stability Investigation of Linear Autonomous NFDEs

All the methods from Chapter 2, §3 may be applied for investigation of characteristic functions corresponding to linear autonomous NFDE. The

§1. Stability of Linear Autonomous NFDEs

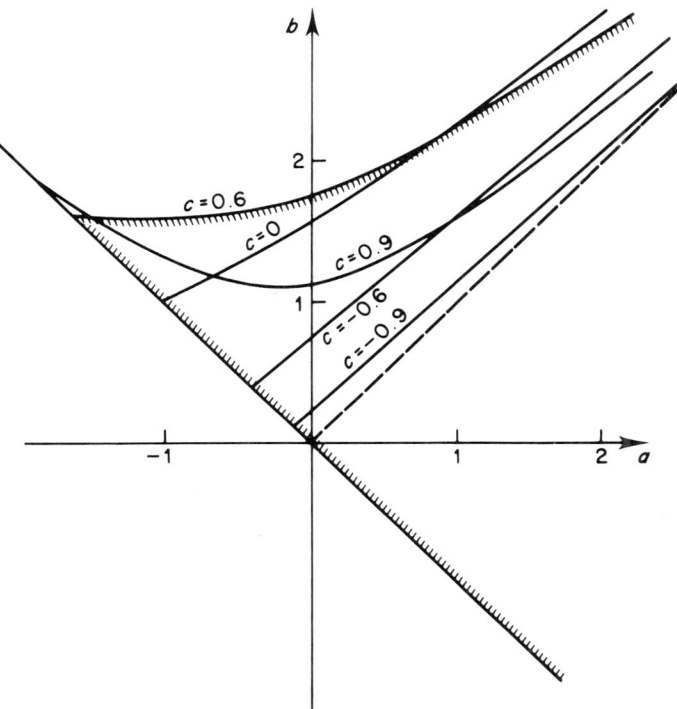

Fig. 3.1. Stability domains for Eq. (1.17).

theorems of Pontrjagin and Chebotarev are valid for NFDEs without any changes and the *method of D subdivision* is also applicable. The stability domains for the equation

$$\dot{x}(t) = ax(t) + bx(t - \tau) + c\dot{x}(t - \tau) \qquad (1.18)$$

in the space of parameters (a, b) for $c = -0.9, -0.6, 0, 0.6, 0.9$ are shown in Fig. 3.1. Equation (1.1) is always unstable for $|c| > 1$ because the corresponding quasi-polynomial has a chain of asymptotic zeros of the form (1.5.5), with the positive real part equal to $\ln|c|$. For all $|c| < 1$, Eqs. (1.1) have a common part of the stability domains $a > |b|$. It follows from the Tsypkin criterion (see Chapter 2, §3).

Vyshnegradskii diagrams for control system (1.6.1), having the transfer function $W(s) = K(Ts + 1)^{-1} \exp(-s\tau)$, closed by a PD controller and by a PID controller in parameters $T_1 = T^{-1}$, $I_1 = I/T$, $D_1 = D/T$, are shown in Figs. 3.2 and 3.3 (consequent equations can be found in Table 1.2).

118 3. Stability of Neutral Functional Differential Equations

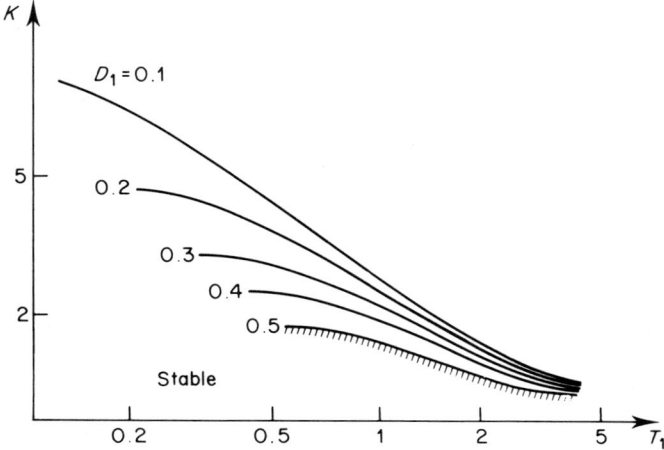

Fig. 3.2. Vyshnegradskii diagram for a system closed by a PD controller.

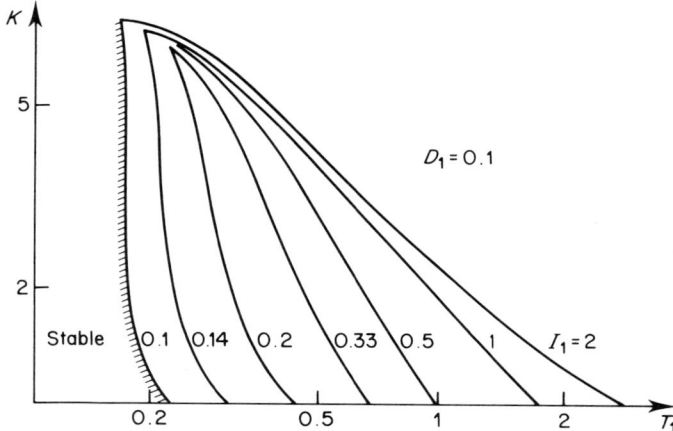

Fig. 3.3. Vyshnegradskii diagram for a system closed by a PID controller.

§1. Stability of Linear Autonomous NFDEs

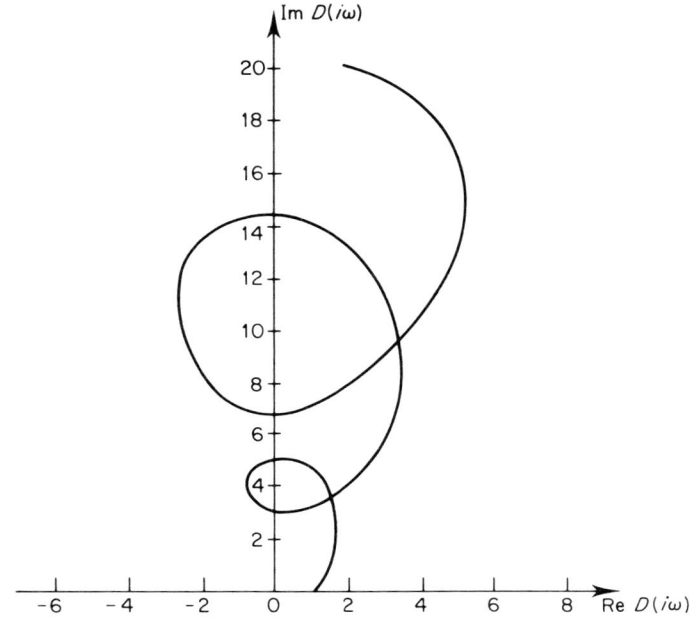

Fig. 3.4. Michailov hodograph for a stable NFDE.

The frequency stability criteria of Michailov and Nyquist from Chapter 2, §3 are not valid for NFDEs in general. It is connected with oscillations in the Michailov hodograph. Consider, for example, the equation

$$\dot{x}(t) = -ax(t) = c\dot{x}(t-1), \qquad a > 0 \tag{1.19}$$

The Michailov hodograph $D(i\omega) = i\omega - a + ci\omega \exp(-i\omega)$, $0 \le \omega < \infty$, for stable Eq. (1.19) when $|c| < 1$ is represented in Fig. 3.4 and for unstable Eq. (1.19) when $|c| > 1$ in Fig. 3.5.

From Figs. 3.4 and 3.5 it follows that the functions arg $D(i\omega)$ have no limit for $\omega \to \infty$ and that their increments on the interval $0 \le \omega < \infty$ are not defined. But Michailov criterion is true [7] for a particular characteristic function

$$\det\left[Iz - H - \int_0^\infty J(s)e^{-zs}\,ds - z\int_0^\infty J_1(s)e^{-zs}\,ds \right]$$

where

$$\int_0^\infty |J(s)|e^{\mu s}\,ds < \infty, \qquad \int_0^\infty |J_1(s)|e^{\mu s}\,ds < \infty, \qquad \mu > 0$$

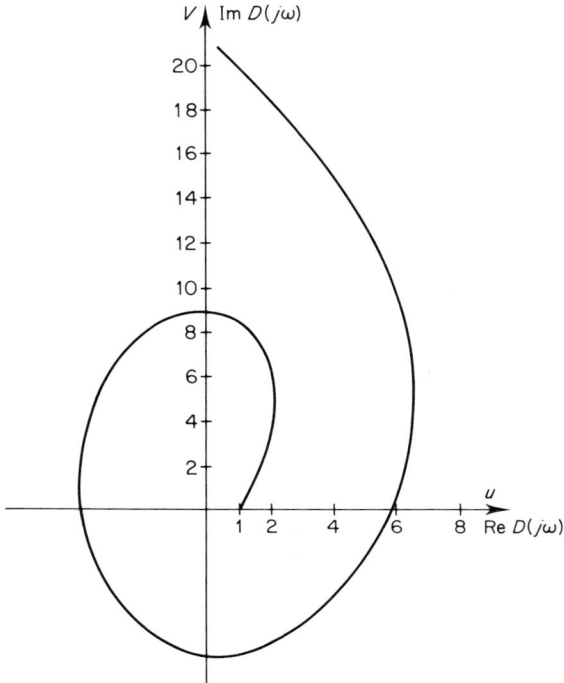

Fig. 3.5. Michailov hodograph for an unstable NFDE.

§2. STABILITY OF AEROAUTOELASTICITY EQUATIONS

2.1. Equations of Unsteady Motion of an Elastic Rigid Body

The dynamic model of unsteady motion of the elastic flying vehicle must take into account the interaction of aerodynamic, elastic and inertial forces (i.e., aeroelastic effects) as well as the dependence of aerodynamic characteristics on the past states of flow (or on the aerodynamic trace). One such description is the aeroautoelasticity models [19, 20].

Practically, one can consider only small perturbations of the motion and deformation parameters. The widely used method of known forms is based on the representation of deformations as a series of known forms which are usually the natural oscillation forms of the vehicle in a vacuum. If one is restricted to a finite number n of the oscillation forms, one can construct an approximate finite-dimensional model of flying vehicle motion

$$\mu(\ddot{q} + 2\omega æ \dot{q} + \omega^2 q) = C \qquad (2.1)$$

§2. Stability of Aeroautoelasticity Equations

Here μ, ω, æ are the diagonal $(n \times n)$ matrices of generalized masses, frequencies and damping factors of natural oscillations; q is a vector of generalized coordinates; and C is a vector of generalized forces. The matrices μ, ω, æ are defined simultaneously with the calculation of natural oscillation forms on the basis of a mass-elastic model of the vehicle. The knowledge of a vector q permits us to calculate the displacements $\bar{p}$ of any point (ξ, η, ζ) of the vehicle and their derivatives according to the expression

$$\bar{p}(\xi, \eta, \zeta) = \sum_{i=1}^{n} q_i(t) \bar{f}_i(\xi, \eta, \xi) \qquad (2.2)$$

Here $\bar{f}_i$ are natural oscillation forms of the vehicle in the coordinate system (ξ, η, ζ) fixed in the principal inertial axes of the vehicle. The vector of the generalized forces C for the known aerodynamic pressure $\bar{p}(\xi, \eta, \zeta, t)$, applied to the vehicle surface σ, is

$$C_i = \int_\sigma \bar{f}_i \bar{p} \, d\sigma, \quad i = 1, \ldots, n, \quad \bar{p} = \sum p_{\varepsilon_j} \qquad (2.3)$$

Here $\varepsilon_j, j = 1, \ldots, N_\varepsilon$ are the kinematic parameters, including vectors $q(t)$ and $\dot{q}(t)$, vector of wind velocity $\Delta(t)$, and also vectors $\delta(t)$, $\dot{\delta}(t)$, describing the states and the rates of states of control surfaces. We write the reaction of component C_i of the generalized force vector on the step change of ε_j (i.e., the unit step function) in the form

$$H_{c_i}^{\varepsilon_j}(\theta) = C_{c_i}^{\varepsilon_j} + I_{c_i}^{\varepsilon_j}(\theta) \qquad (2.4)$$

Hence $C_c^{\varepsilon_j}$ is a *quasi-stationary aerodynamic derivative*. The functions $I_{c_i}^{\varepsilon_j}(\theta)$ are called the *unsteady unit step functions*. They describe the influence of the aerodynamic trace. For the subsonic speed (Mach number $M < 1$) the function $I_{c_i}^{\varepsilon_j}(\theta)$ are determined for all moments θ, $0 \leq \theta < \infty$ and for sufficiently great θ, $I_{c_i}^{\varepsilon_j}(\theta) \cong A\theta^{-2}$. For the supersonic speed ($M > 1$) $I_{c_i}^{\varepsilon_j}(\theta) = 0$ for $\theta \geq \theta^* > 0$ [7, 19]. The unsteady unit step functions are calculated theoretically on the base of the discrete vortices theory or determined experimentally. Usually they are represented by tables or by diagrams. Figure 3.6 shows the unit step function of aerodynamic moment m_z with respect to the diametrical axis, passing by the choice of a rectangular wing with aspect ratio of $\lambda = 5$. Curve 1 corresponds to $M = 0.7$ and curve 2 to $M = 0.9$.

The quasi-stationary values of moment $m_z = H_{m_z}^\alpha(\infty)$ are indicated by dotted lines on Fig. 3.6. Using the Duhamel integral, we can write any aerodynamic characteristic in the form

$$C_i(t) = \sum_j C_{\varepsilon_j}, \quad C_{\varepsilon_j}(t) = C^{\varepsilon_j}\varepsilon_j + \int_0^t I_{C_i}^{\varepsilon_j}(t - \theta)\dot{\varepsilon}(\theta) \, d\theta \qquad (2.5)$$

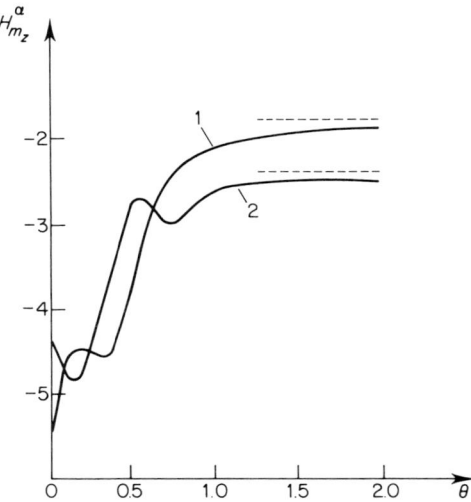

Fig. 3.6. Unsteady unit step function of aerodynamic moment m_z: 1, $M = 0.7$; 2, $M = 0.9$.

Substituting (2.5) in (2.1) we obtain the *closed linear equations of flying vehicle perturbated motion*

$$\mu(\ddot{q} + \omega \alpha \dot{q} + \omega^2 q) - C^q q - C^{\dot{q}} \dot{q} - \int_0^t I_C^q(\theta) \dot{q}(t - \theta) \, d\theta$$

$$- \int_0^t I_C^{\dot{q}}(\theta) \ddot{q}(t - \theta) \, d\theta = C^\delta \delta + C^{\dot{\delta}} \dot{\delta} + C^\Delta \Delta$$

$$+ \int_0^t I_C^\delta(\theta) \dot{\delta}(t - \theta) \, d\theta + \int_0^t I_C^{\dot{\delta}}(\theta) \ddot{\delta}(t - \theta) \, d\theta$$

$$+ \int_0^t I_C^\Delta(\theta) \dot{\Delta}(t - \theta) \, d\theta \qquad (2.6)$$

Equations (2.6) describe the deterministic model of flying vehicle dynamics. The stochastic model is more adequate. It takes into account the atmospheric turbulences and chaotic noises in the pickups. The stochastic differential equations for the turbulent wind velocity $\dot{\Delta}$ and the model of pickups are taken in the form

$$\dot{\Delta} = F_\Delta \Delta + \dot{v}_1, \qquad \dot{r} = Mr + \hat{x} + \dot{v}_2$$

Here $\dot{v}_1, \dot{v}_2$ are vector stochastic processes of the white noise type with known intensity matrices, r is a vector of pickup states and $\hat{x}$ is a vector of output

§2. Stability of Aeroautoelasticity Equations

pickup signals. The drives of control surfaces are described by the equation $\dot{\delta} = S\delta + Gu$ where u is a vector of control drives signals. Finally, the *linear stochastic aeroautoelastic model* has the form

$$\dot{x}(t) = Fx(t) + \int_0^t I_0(\theta)x(t-\theta)\, d\theta$$
$$+ \int_0^t I_1(\theta)\dot{x}(t-\theta)\, d\theta + Gu(t) + \dot{v} \qquad (2.7)$$

Equation (2.7) is an SFDE of the neutral type. In Chapter 4 we shall formulate the theorem of existence and uniqueness of the solutions of such equations and shall give some methods of stability investigation.

2.2. Stability of Aeroautoelastic Equations

Study the stability of deterministic equation (2.6) for the control vector $u \equiv 0$:

$$\ddot{q}(t) + A\dot{q}(t) + Bq(t) = \int_0^t I_0(\theta)\dot{q}(t-\theta)\, d\theta$$
$$+ \int_0^t I_1(\theta)\ddot{q}(t-\theta)\, d\theta \qquad (2.8)$$

According to Theorem 1.2 for asymptotic stability of the solution of Eq. (2.8), it is necessary and sufficient that the characteristic function $D(z)$ has no zeros in the half-plane Re $z \geq 0$. The characteristic function $D(z)$ is defined with the aid of the theorem on the Laplace transform of convolution and is equal to

$$D(z) = \det[Iz^2 + Az + B - z\int_0^\infty I_0(s)e^{-zs}\, ds - z^2 \int_0^\infty I_1(s)e^{-zs}\, ds] \qquad (2.9)$$

The Michailov criterion holds for function (2.9) if for some $\gamma > 0$

$$\int_0^\infty I_i(s)e^{-\gamma s}\, ds < \infty, \qquad i = 0, 1$$

The kernels $I_i(s)$ are usually uncertainties $M_i(s)$, $i = 0, 1$. It is shown by Astapov [7] that if function (2.9) has no zeros in the half-plane Re $z \geq -\lambda$, $\lambda > 0$ and the value

$$\|M_1(0)\| + \int_0^\infty \left\|M_0(\theta) + \frac{d}{d\theta}M_1(\theta)\right\| d\theta$$

is sufficiently small then the replacement of kernels $I_i(\theta)$ by $I_i(\theta) + M_i(\theta)$, $i = 0, 1$, does not distort the asymptotic stability. Consider in detail the

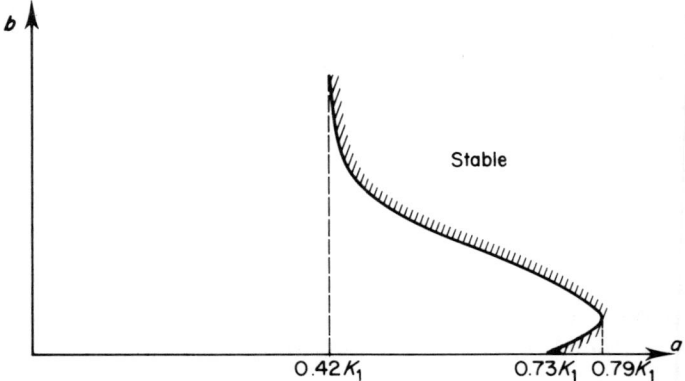

Fig. 3.7. Stability domain of triangular wing for $M = 1.2$.

one-dimensional oscillation of the control surface, the so called control surface buzz [121]. Let a rigid wing move in gas flow and turn with respect to an axis. The angular wing displacements are restricted by an elastic spring of rigidity k. The *motion equation* is

$$\ddot{q} + a\dot{q} + bq = k_1 \int_0^\infty I_0(\theta)\dot{q}(t-\theta)\, d\theta + k_1 \int_0^t I_1(\theta)\ddot{q}(t-\theta)\, d\theta$$

$$a = -k_1 m_z^{\omega z}, \qquad b = k_2 - k_1 m_z^\alpha, \qquad k_1 = \rho S \bar{b}^{-3}/(2J_z) \quad (2.10)$$

$$k_2 = kb^2/(J_z u_0^2)$$

If one considers the quasi-stationary model (i.e., $J_0(\theta) = 0$, $J_1(\theta) = 0$), then Eq. (2.10) is asymptotically stable for $a > 0$, $b > 0$. Investigate the stability of a triangular wing with aspect ratio $\lambda = 2.5$ for $M = 1.2$ and $M = 2$. We shall use the results of Astapov [7], and Belotzerkovskii [19], who give the aerodynamic coefficients and unsteady step functions. Figure 3.7 presents the stability regions obtained with the aid of the Michailov criterion for these cases. Figures 3.8 and 3.9 present the stability region of a rectangular wing with $\lambda = 5$ and $\lambda = 2.5$ for $M = 0.4$. It is clear from these figures that aerodynamic aftereffects increase the stability region for subsonic speed and decrease it for supersonic speed. The new qualitative phenomenon is also revealed. The vertical straight line crosses the stability region boundary twice: on Fig. 3.7 for $M = 1.2$ and $0.73k_1 < a < 0.79k_1$, and on Fig. 3.8 for $-0.76k_1 < a < -0.64k_1$. It means that the wing is stable if the speed is less than a critical speed u_{cr}^1 or more than the other critical speed u_{cr}^2. For the speeds varying from u_{cr}^1 to u_{cr}^2 the wing is unstable; it oscillates with increasing amplitude and after some time goes to ruin. This effect is observed in practice.

§2. Stability of Aeroautoelasticity Equations

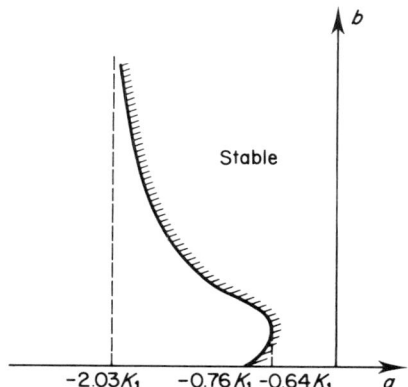

Fig. 3.8. Stability domain of rectangular wing for $\lambda = 5$ and $M = 0.4$.

It can be explained in the models which take into account the aerodynamic aftereffects.

The *two-dimensional flexure-torsion flutter* of a rigid rectangular wing is considered by Astapov [7], and Belotzerkosvii [19]. The critical flutter speeds are obtained: in the model with aftereffects it is equal to 290 m/sec, and in the quasi-stationary model it is equal to 190 m/sec. Thus, allowing for the aftereffects of the aerodynamic trace permits us to obtain more adequate models of unsteady motion (aeroautoelasticity model) which gives more precise quantitative and qualitative descriptions of the unsteady motion of an elastic flying vehicle.

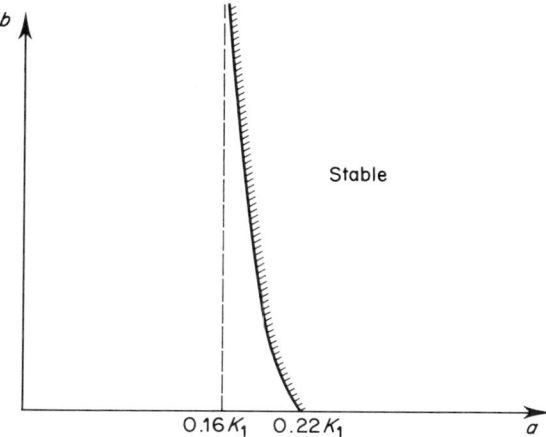

Fig. 3.9. Stability domain of rectangular wing for $\lambda = 2.5$ and $M = 0.4$.

§3. LIAPUNOV DIRECT METHOD FOR NEUTRAL-TYPE EQUATIONS

3.1. Introduction

Application of the Liapunov direct method to NFDEs is a nontrivial problem. Several assertions of this type could be obtained by formal extension of the corresponding theorems for equations with delay. It should be noted however that the results thus obtained assuming the existence of positive definite Liapunov functionals of systems' paths have limited applications. The reason is that it is very difficult to construct Liapunov positive-definite functionals for NFDEs. Instead of this, another method of stability investigation is stated below based on the use of only positive-semidefinite Liapunov functionals. In this case the investigation consists of two stages. At the first stage a positive-semidefinite (but not, as usual, positive-definite!) functional is constructed. The second stage is connected with the study of stability of some functional inequalities generated by the positive-semidefinite functional constructed earlier. Some results of this section are established for equations with arbitrary (finite or infinite) aftereffect. The considered method of stability investigation of NFDE with the help of positive-semidefinite functionals was suggested by Kolmanovskii and Nosov [108(1)]. In Kolmanovskii and Nosov [108(2), 108(5)] positive-semidefinite functionals were built for concrete systems and the stability conditions formulated in terms of the systems' coefficients were obtained. Some problems connected with the application of the Liapunov direct method to NFDEs are considered in other works [48, 81(4), 150, 179].

3.2. Some Definitions

Denote M as the metric space of continuous functions $\varphi(t)$, $\varphi: (-\infty, 0] \to R_n$ and Q_H is a sphere in M (see Chapter 1, §3). Let $F, G: [0, \infty] \times Q_H \to R_H$ be two preassigned continuous maps such that

$$|G(t, \varphi)| + |F(t, \varphi)| \le F_0, \quad t \in [0, \infty), \quad \varphi(\theta) \in Q_H \quad (3.1)$$

Consider the initial-value problem for the NFDE

$$(d/dt)[x(t) - G(t, x_t)] = F(t, x_t), \quad t \ge 0 \quad (3.2)$$

$$x_0(\theta) = \varphi(\theta), \quad -\infty < \theta \le 0, \quad x_t = x(t + \theta) \in M \quad (3.3)$$

Assume hereafter that the solutions of problem (3.2), (3.3) exists, is unique and continuously depends on initial data. Let $G(t, 0) = F(t, 0) = 0$. Then system (3.2) has a trivial solution corresponding to the initial function $\varphi(\theta) = 0$.

§3. Liapunov Direct Method for Neutral-Type Equations

Definition 3.1. *The trivial solution of problem* (3.2), (3.3) *is called* (a) *stable if for any* $\varepsilon > 0$ *there exists* $\delta(\varepsilon) > 0$ *such that* $|x(t, \varphi)| \leq \varepsilon$ *for* $t \geq 0$ *as soon as* $\rho(\varphi, 0) \leq \delta(\varepsilon)$; (b) *asymptotically stable if it is stable and*

$$\lim_{t \to \infty} x(t, \varphi) = 0, \quad \forall\; \varphi(\theta) \in \Omega \subset M$$

The region Ω is referred as the attraction region of the trivial solution.

Later, a great role will be played by *functional inequalities*

$$|Z(t, y_t)| = |y(t) - G(t, y_t)| \leq f(t), \quad y_0(\theta) = \varphi(\theta) \tag{3.4}$$

Here $f(t)$ is a nonnegative continuous scalar function and $\varphi(\theta) \in M$. Designate by $y(t, \varphi)$ any solution of inequality (3.4).

Definition 3.2. *The trivial solution* $y(t) = 0$ *of inequality* (3.4) *is called* (a) f *stable if for any* $\varepsilon > 0$ *there exists* $\delta(\varepsilon) > 0$ *such that* $|y(t, \varphi)| \leq \varepsilon$, $t \geq 0$, *for all φ and f such that* $\rho(\varphi, 0) \leq \delta(\varepsilon)$, $\sup_t f(t) \leq \delta(\varepsilon)$; (b) *asymptotically f stable if it is f stable and moreover* $\lim_{t \to \infty} y(t, \varphi) = 0$ *for all* $\varphi(\theta) \in \Omega \subset M$ *and any $f(t)$ such that $f(t) \to 0$ at $t \to \infty$.*

3.3. General Theorems

Let $V(t, x_t, Z(t, x_t))$ be some continuous functional determined for all $x_t \in Q_H$ and $t \geq 0$ and such that its derivative $\dot{V} = dV(t, x_t, Z(t, x_t))/dt$ exists along the trajectory of equation (3.2). Since the derivative $dZ(t, x_t)/dt$ exists and the derivative $\dot{x}(t)$ may not exist, the requirement of the existence of the derivative $\dot{V}$ imposes certain restrictions on the dependence of V on x_t. As in Chapter 2, §5, we designate by $\omega_i(u)$ some continuous nondecreasing functions such that $\omega_i(0) = 0$ and $\omega_i(u) > 0$ for $u > 0$.

Theorem 3.1. *Let the functional* $V(t, x_t, Z(t, x_t))$ *exist and satisfy the previously mentioned assumptions. Further,*

$$\omega_1(|Z(t, x_t)|) \leq V(t, x_t, Z(t, x_t)) \leq \omega_2(\rho(x_t, 0)) \tag{3.5}$$

$$\dot{V} \leq 0 \tag{3.6}$$

Let the trivial solution of the inequality (3.4) *be f stable. Then the trivial solution of equation* (3.2) *is stable.*

Proof. Take an arbitrary $\varepsilon \in (0, H)$. Since the inequality (3.4) solution $g(t) = 0$ is f stable, one can find $\delta_1(\varepsilon) > 0$ such that any solution of inequality (3.4) will satisfy the relation $|x(t, \varphi)| \leq \varepsilon$, $t \geq 0$, if $f(t) \leq \delta_1$, $\rho(\varphi, 0) \leq \delta_1$. Now take $\delta_2 \in (0, \delta_1)$ such that $\omega_2(\delta_2) = \omega_1(\delta_1)$. Then $\rho(\varphi, 0) \leq \delta_2$ by virtue of conditions (3.5) and (3.6) we obtain $\omega_1(|Z(t, x_t)|) \leq V(t, x_t, Z(t, x_t)) \leq V(0, \varphi, Z(0, \varphi)) \leq \omega_2(\delta_2) = \omega_1(\delta_1)$. This and the monotonocity of ω_1 imply that $|Z(t, x_t)| \leq f(t) \leq \delta_1$. Taking into account the f stability of inequality (3.4), we obtain $|x(t, \varphi)| \leq \varepsilon$, $t \geq 0$, for all $\varphi(\theta)$ such that $\rho(\varphi, 0) \leq \delta_2 \leq \delta_1$. Theorem 3.1 is proved.

Theorem 3.2 [108(5)]. *Let there exist a functional $V(t, x_t, Z(t, x_t))$ satisfying requirements of Theorem* 3.1 *and also*

$$\dot{V} \leq -\omega_3(|Z(t, x_t)|) \tag{3.7}$$

Let the trivial solution of inequality (3.4) *be asymptotically f stable. Then the trivial solution of equation* (3.2) *is asymptotically stable.*

As a corollary, consider the ordinary differential equation

$$\dot{x}(t) = F(t, x(t)), \qquad x(t_0) = x_0 \tag{3.8}$$

Here $F: [0, \infty) \times R_n \to R_n$ is a continuous function satisfying the local Lipschitz condition in the second argument, and $F(t, 0) = 0$. Along with (3.8) consider an NFDE

$$\begin{aligned}\dot{Z}(t, x_t) &= F(t, Z(t, x_t)), \qquad x_0(\theta) = \varphi(\theta) \\ Z(t, x_t) &= x(t) - G(t, x_t)\end{aligned} \tag{3.9}$$

Here $G(t, x_t)$ satisfies the conditions formulated previously.

Theorem 3.3. *Let the trivial solution of Eq.* (3.8) *be asymptotically stable. Further, let the trivial solution of inequality* (3.4) *be asymptotically f stable. Then the trivial solution of* (3.9) *is asymptotically stable.*

For the proofs of Theorems 3.2 and 3.3 see Kolmanovskii and Nosov [108(5)].

3.4. Global Stability

Investigate the problem of global stability of the trivial solution of Eq. (3.2). Assume that the functionals $F(t, \varphi)$ and $G(t, \varphi)$ are continuous and determined on the whole space $[0, \infty) \times M$. In addition, $|F(t, \varphi)| \leq F_H$, $|G(t, \varphi)| \leq G_H$ for all $H > 0$, $\varphi \in Q_H$ and $t \geq 0$. Suppose also that the conditions of existence and uniqueness are fulfilled for an arbitrary sphere S_H.

Definition 3.3. *The trivial solution of Eq.* (3.2) *is called globally asymptotically stable if it is stable and* $\lim x(t, \varphi) = 0$, $t \to \infty$ *for any initial function* $\varphi(\theta) \in M$.

Definition 3.4. *The trivial solution of inequality* (3.4) *is called f bounded if bounded solution $y(t, \varphi)$ corresponds to every bounded function $f(t)$.*

§3. Liapunov Direct Method for Neutral-Type Equations

Theorem 3.4. *Let the continuous functional $V(t, x_t, Z(t, x_t))$ exist and satisfy all the conditions of Theorems 3.2. In addition, let the trivial solution of inequality (3.4) be f bounded and asymptotically f stable, and also let*

$$\omega_1(u) \to \infty, \quad u \to \infty \tag{3.10}$$

Then the trivial solution of Eq. (3.2) is globally asymptotically stable.

The proof of this theorem may be obtained by suitable modification of the results from Kolmanovskii and Nosov [108(5)].

3.5. Stability of the Functional Inequalities

Establish some conditions of f stability for the case in which the space M coincides with $CB[-\infty, 0]$.

Lemma 3.1. *Let the continuous functional $G(t, \varphi)$, defined on $[0, \infty) \times CB[-\infty, 0]$, satisfy the Lipschitz condition*

$$|G(t, \varphi) - G(t, \psi)| \le v\|\varphi - \psi\|_B \tag{3.11}$$

If, in addition,

$$v < 1 \tag{3.12}$$

then the trivial solution of inequality (3.4) is f stable and f bounded.

Proof. From (3.4) it follows that

$$|y(t)| \le |G(t, y_t)| + f(t) \le v\|y_t(\theta)\|_B + f(t)$$

Designate $m(t) = \sup_{0 \le s \le t} |y(s)|$. Then

$$m(t) \le vm(t) + v\|\varphi(\theta)\|_B + \sup_{0 \le s \le t} f(s) \tag{3.13}$$

The estimate (3.13) means that the trivial solution of inequality (3.4) is f stable and f bounded.

Formulate some assertions [108(5)] about asymptotic f stability of the trivial solution of inequality (3.4) in the case of a finite time lag.

Lemma 3.2. *Let $G(t, y_t) = g(t, y(t - h))$, where $g: R_1 \times R_n \to R_n$ and $h > 0$. Then if $|g(t, y(t - h))| \le v|y(t - h)|$, $0 < v < 1$, then the trivial solution of the inequality $|y(t) - g(t, y(t - h))| \le f(t)$ is asymptotically f stable and f bounded.*

Recall some concepts connected with the theory of almost-periodic functions. The spectrum of the almost-periodic function $\varphi(t)$ is the set $\Lambda(\varphi)$ such that

$$\Lambda(\varphi) = \{\lambda: R_1 : M[e^{i\lambda t}\varphi(t)] \ne 0\}$$

$$M[\psi] = \lim_{T \to \infty} \frac{1}{T} \int_0^T \psi(t)\, dt$$

The module mod(φ) of the almost-periodic function φ is the set

$$\mathrm{mod}(\varphi) = \left\{ \sum_{i=1}^{K} m_i \lambda_i, \ \lambda_i \in \Lambda(\varphi), \ m_i \text{ an integer}, \ K \text{ a natural number} \right\}$$

Lemma 3.3. *Let* $G(t, y_t) = g(t, y(t - h))$, $g: R_1 \times R_n \to R_n$, $h > 0$, $|g(t, y(t - h))| \le v(t)|y(t - h)|$, $v(t) > 0$. *Let the continuous function* $v(t)$ *be almost periodic and satisfy the conditions*

$$\forall \{\lambda: \lambda \in \mathrm{mod}(v), \ \lambda \ne 0, \ [h\lambda \ne 0(\mathrm{mod}\ 2\pi)]\}, \qquad M[\ln v(t)] < 0$$

Either $v(t)$ *is* ω *periodic, h is incommensurable with* ω *and*

$$\int_0^\omega \ln v(t)\, dt < 0$$

Then the trivial solution of (3.4) is asymptotically f stable and f bounded.

Lemma 3.4. *Let the functional* $G(t, \varphi)$ *satisfy the Lipschitz condition*

$$|G(t, \varphi) - G(t, \psi)| \le \alpha \|\varphi - \psi\|, \ 0 < \alpha < \frac{1}{2}$$

and also $G(t, \varphi)$ *is independent from the values of the function* $\varphi(\theta)$, $\theta \in [-\Delta, 0]$. *Here* $0 < \Delta < h$. *Then the trivial solution of (3.4) is asymptotically f stable and f bounded.*

3.6. Equations with Bounded Delay

Formulate the stability conditions for the case of Eqs. (3.2) with finite delay and for those with infinite delay such that the initial set coincides with the bounded interval $[-h, 0]$. So F is a continuous functional $F: [0, \infty) \times C[-h, t] \to R_n$.

Theorem 3.5. *Assume that the functional* $G(t, \varphi)$, $G: [0, \infty) \times C[-h, 0] \to R_n$ *satisfies the condition*

$$|G(t + s, \varphi) - G(t, \psi)| \le \omega_5(s) + \alpha \|\varphi - \psi\|$$

$$0 < \alpha < 1, \quad \forall \varphi, \psi \in Q_H, \quad 0 \le t < \infty, \quad s \ge 0 \qquad (3.14)$$

Let there exist the functional $V(t, x_t, Z(t, x_t))$ *satisfying condition (3.5) and*

$$\dot{V} \le -\omega_6(|x(t)|) \qquad (3.15)$$

Then the trivial solution of (3.1) is asymptotically stable.

§3. Liapunov Direct Method for Neutral-Type Equations

Proof. By virtue of Theorem 3.1 and Lemma 3.1 the trivial solution of (3.2) is stable. Hence $x(t, \varphi) \in Q_H$ for $\varphi(\theta) \in Q_\delta$, $\delta > 0$. Show that $x(t, \varphi) \to 0$, $t \to \infty$. Suppose the contrary. Then there exist such a number $v > 0$ and a sequence of points $t_i \to \infty$ such that $|x(t_i)| > v$. From (3.2) it follows that

$$x(t) = G(t, x_t) + \int_0^t F(s, x_s)\, ds$$

Here it is assumed that $x(s) = \varphi(s)$ for $s \leq 0$. Then taking into account (3.1), we have for $\Delta \geq 0$

$$|x(t + \Delta) - x(t)|$$
$$= |G(t + \Delta, x_{t+\Delta}) - G(t, x_t)$$
$$+ \int_t^{t+\Delta} F(s, x_s)\, ds|$$
$$\leq \alpha \|x_{t+\Delta} - x_t\| + \omega_5(\Delta) + F_0 \Delta \qquad (3.16)$$

Designate

$$\rho(\eta) = \sup_{t \geq 0} \sup_{\Delta \leq \eta} |x(t + \Delta) - x(t)|$$

From (3.16) it follows that

$$\rho(\eta) \leq (1 - \alpha)^{-1} \left[\alpha \max_{\eta \geq \Delta} \max_{-h \leq \theta \leq 0} |x(\theta + \Delta) - \varphi(\theta)| + \omega_5(\eta) + F_0 \eta \right]$$

So, due to the uniform continuity of the function $x(t)$ on the closed segment $[-h, \eta]$, the right-hand side of this inequality tends to zero as $\eta \to 0$. Hence there exists a $\bar{\eta} > 0$ such that $\rho(\bar{\eta}) \leq v/2$. In this case for $\tau \in [t_i - \bar{\eta}, t_i + \bar{\eta}]$ we have $|x(\tau)| > v/2$ uniformly for all i. From this follows a contradiction: $V \to \infty$ for $t \to \infty$. Theorem 3.5 is proved

REMARK 3.1. The assertion of Theorem 3.5 is valid for the functional $G(t, x_t)$, $G: [0, \infty) \times C[-h, t] \to R_n$ satisfying the condition

$$|G(t + s, x_{t+s}) - G(t, x_t)| \leq \omega_5(s) + \alpha \max_{-h \leq \tau \leq t} |x(\tau + s) - x(\tau)|$$

$$0 < \alpha < 1, \quad s \geq 0, \quad \forall\, x_t \in Q_H, \quad t \geq 0$$

3.7. Examples

EXAMPLE 3.1. Consider an equation of the type (3.2)

$$\ddot{x} + C\ddot{x}(t - h) + g(x(t) + Cx(t - h)) = 0, \quad h > 0 \qquad (3.17)$$

where $xg(x) > 0$ for $x \neq 0$ and $|C| < 1$. Prove that the trivial solution of (3.17) will be stable. Write (3.17) in the form

$$Z(t, x_t) = x(t) + Cx(t - h), \qquad \dot{Z} = W, \qquad \dot{W} = -g(Z)$$

Consider the functional

$$V = \frac{W^2}{2} + \int_0^Z g(s)\,ds \tag{3.18}$$

The derivative of functional (3.18) is equal to

$$\dot{V} = W\dot{W} + \dot{Z}g(Z) = 0$$

Hence by virtue of Theorem 3.1 and Lemma 3.1 the trivial solution of Equation (3.17) is asymptotically stable.

EXAMPLE 3.2. Derive the stability conditions for the equations

$$\ddot{x}(t) + \varphi(x(t))\dot{x}(t) + f(x) = 0, \qquad f(0) = 0 \tag{3.19}$$

$$\ddot{Z}(t, x_t) + \varphi(Z(t, x_t))\dot{Z}(t, x_t) + f(Z(t, x_t)) = 0 \tag{3.20}$$

where $Z(t, x_t) = x(t) + \frac{1}{2}[\exp(\sin t)]x(t - 1)$. Under the conditions $xf(x) > 0$, $x \neq 0$, $\varphi(x) > 0$, the trivial solution of Eq. (3.19) is uniformly asymptotically stable [15]. By virtue of Theorem 3.3 and Lemma 3.3 the trivial solution of Eq. (3.20) is also asymptotically stable. If, in addition,

$$\int_0^x f(s)\,ds \to \infty, \qquad x \to \infty$$

then the trivial solution of Eq. (3.20) is globally asymptotically stable.

§4. CONSTRUCTION OF DEGENERATED FUNCTIONALS FOR CONCRETE SYSTEMS

4.1. Stability of a Chemical Reactor Closed by a PD Controller

Investigate the stability of the chemical reactor described in Chapter 1, §1 and Chapter 2, §6. Figure 3.10 represents the structural scheme of the investigated system. As in Chapter 2, §6 the output $u(t)$ of a nonlinear actuator is considered as a functional depending on the preceding states of the input $x(s)$, $0 \leq s \leq t$, i.e., $u(t) = F(x(t + \theta))$, $-t \leq \theta \leq 0$. Assume also that this functional satisfies Lipschitz condition (2.6.5) from Chapter 2, §6.

§4. Construction of Degenerated Functionals for Concrete Systems

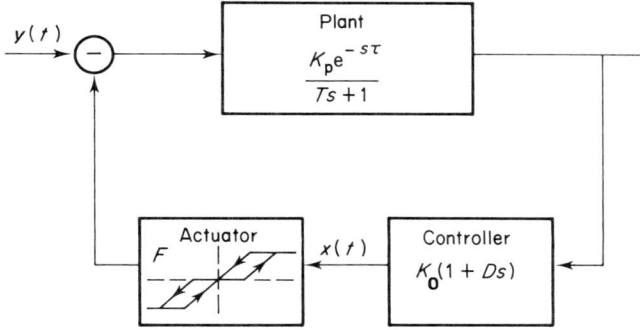

Fig. 3.10. Structural scheme of a chemical reactor closed by a PD controller.

The functioning of the system illustrated in Fig. 3.10 is described by the equation

$$(d/dt)[Tx(t) + K_0 K_p DF(x(t - \tau + \theta))]$$
$$= -x(t) + K_0 K_p F(x(t - \tau + \theta)) + K_0 K_p [D\dot{y}(t - \tau) + y(t - \tau)]$$

Let $\bar{y}(t)$ be some fixed input signal, and let $\bar{x}(t)$ be the unperturbed solution. Designate the perturbed solution by $\bar{x}(t) + z(t)$. The perturbation $z(t)$ satisfies the equation

$$(d/dt)[Tz(t) + K_0 K_p DF_1(z(t - \tau + \theta))]$$
$$= -z(t) + K_0 K_p F_1(z(t - \tau + \theta)) \tag{4.1}$$

$$F(z(t - \tau + \theta))$$
$$= F(\bar{x}(t - \tau + \theta) + z(t - \tau + \theta)) - F(\bar{x}(t - \tau + \theta))$$

Introduce the following functional

$$V(t, z_t) = [Tz(t) + K_0 K_p DF_1(z(t - \tau + \theta))]^2$$
$$+ \gamma \int_0^\infty d_s R_1(s, \bar{x}_s, z_s) \int_{t-\tau-s}^t z^2(s_1)\, ds_1 \tag{4.2}$$

The derivatives of the functional (4.2) is equal to

$$\dot V = 2[Tz(t) + K_0 K_p DF_1(z(t - \tau + \theta))][-z(t) + K_0 K_p F_1(z(t - \tau + \theta))]$$
$$+ \gamma z^2(t) \int_0^\infty d_s R(s, \bar x_s, z_s) - \gamma \int_0^\infty z^2(t - \tau - s) d_s R(s, \bar x_s, z_s)$$
$$\leq -Tz^2(t) + K_0 K_p(D + T)\frac{1}{\varepsilon} z^2(t) + \gamma r^2 z^2(t)$$
$$+ 2K_0^2 K_p^2 DF_1^2(z(t - \tau + \theta)) + K_0 K_p(D + T)\varepsilon F_1^2(z(t - \tau + \theta))$$
$$- \gamma \int_0^\infty z^2(t - \tau - s) d_s R(s, \bar x_s, z_s)$$

Set $\varepsilon = r^{-1}$, $\gamma = K_0 K_p(D + T)r^{-1} + 2K_0^2 K_p^2 D$. In this case we obtain

$$\dot V \leq [-2T + 2K_0 K_p(D + T)r + 2K_0^2 K_p^2 Dr^2]z^2(t)$$

Using Theorem 3.4, we see that the feedback system shown in Fig. 3.10 is asymptotically stable provided that the requirements

$$T > K_0 K_p Dr^2, \qquad 2T > 2K_0 K_p(D + T)r + K_0^2 K_p^2 Dr^2$$

hold. If $D = 0$, then these requirements coincide with stability conditions of a reactor closed by a P controller, obtained in Chapter 2, §6.2.

4.2. Stability of One-Dimensional Nonlinear Systems

Here we investigate the conditions of stability of trivial solutions of scalar equations

$$\dot x(t) = -\int_0^\infty x(t - s)\, dk_0(s) + \int_0^\infty \dot x(t - s)\, dk_1(s) + a(t, x_t)$$
$$t \geq 0, \quad x_t = x(t + \theta), \quad -\infty < \theta \leq 0 \qquad (4.3)$$

In some situations system (4.3) is a particular case of equation (3.2). Therefore, in these situations, from the results of Chapter 3, §3 one can extract some conditions of stability of system (4.3). However, owing to the specific features of Eq. (4.3), one can obtain the stability conditions under wider assumptions. These assumptions are connected first of all with the understanding of the solution of Eq. (4.3) and also with the requirement that the derivative of the Liapunov functional be negative only almost everywhere (but not everywhere, as in Chapter 3, §3). At the same time it should be emphasized that the method of stability investigation of Eq. (4.3) is the same as in Chapter 3, §3.

Formulate the basic assumptions. The kernel $k_0(s)$ has bounded variation

§4. Construction of Degenerated Functionals for Concrete Systems

on $[0, \infty)$, and the corresponding integral in (4.3) is understood in the sense of Stieltjes. The kernel $k_1(s)$ is determined by

$$k_1(s) = \int_0^s \lambda(t)\, dt + \sum_{h_n \leq s} \mu_n, \qquad k_1(0) = 0$$

Here summation extends to those values of n for which $h_n \leq s$. The function $\lambda(s)$ is bounded and Riemann integrable on $[0, \infty)$. Finally, $h_n \geq 0$ and $|\mu_1| + |\mu_2| + \cdots + |\mu_n| + \cdots < \infty$. The functional $a(t, \varphi)$ is defined and continuous on $[0, \infty) \times \mathrm{CB}[-\infty, 0]$. Also, the functional $a(t, \varphi)$ satisfies, for any function $\varphi \in \mathrm{CB}[-\infty, 0]$, the conditions

$$a(t, 0) = 0, \qquad |a(t, \varphi)|^2 \leq \int_0^\infty |\varphi(-s)|^2\, dR_1(s, \varphi)$$

$$\int_0^\infty dR_1(s, \varphi) \leq r_1^2 < \infty, \qquad \forall\, \varphi \in \mathrm{CB}[-\infty, 0]$$

Set

$$\alpha_{ij} = \int_0^\infty s^i |dk_j(s)|$$

It is assumed everywhere below that

$$\alpha_{01} < 1, \qquad \alpha_{00} + \int_0^\infty s\, dR_1(s, \varphi) < \infty \tag{4.4}$$

The solution $x(t)$ of Eq. (4.3) is determined by the initial data

$$x_0(\theta) = \varphi(\theta), \qquad \dot{x}(\theta) = \dot{\varphi}(\theta), \qquad i \leq 0 \tag{4.5}$$

Here $\varphi(\theta)$, $\theta \leq 0$ is an absolutely continuous bounded function, and the function $\dot{\varphi}(\theta)$ is bounded. Under these assumptions there exists the only solution $x(t, \varphi)$ of problem (4.3), (4.5); i.e., there exists the only function $x(t)$ limited and Riemann integrable on each finite interval, which is equal to $\dot{\varphi}(\theta)$ at $\theta \leq 0$ and such that the function $\dot{x}(t)$ and $x(t) = \varphi(0) + \int_0^t \dot{x}(s)\, ds$, $t \geq 0$, will be the solution of problem (4.3), (4.5).

Set for each function $\varphi \in \mathrm{CB}[-\infty, 0]$

$$Z(\varphi) = \varphi(0) - \int_0^\infty \varphi(-s)\, dk_1(s)$$

Notice that when requirements (4.4) are met, the trivial solution of the inequality $|Z(\varphi)| \leq C_0$ is f stable (see Chapter 3, §3).

By analogy with the proof of Theorem 3.5 we establish Theorem 4.1.

Theorem 4.1. *The trivial solution of Eq. (4.3) is globally asymptotically stable, if conditions (4.4) are fulfilled and there exists a functional $V(\varphi)$,*

$$V(\varphi) = W(\varphi) + Z^2(\varphi) \tag{4.6}$$

which satisfies the local Lipschitz conditions, and
$$0 \le W(\varphi) \le \omega_2(\|\varphi\|_B) \tag{4.7}$$
The derivative of functional (4.6) along the trajectories of system (4.3), exists almost everywhere for all $t \ge 0$ and
$$\dot{V} \le -\omega_3(|x(t)|)$$
The proof of Theorem 4.1 can be found in Kolmanovskii and Nosov [108(2), 108(5)].

Obtain on the basis of Theorem 4.1 some concrete stability conditions. Suppose, first, that the kernel $k_0(s)$ has a jump in zero, equal to $a_0 > 0$, and
$$a_0(1 - \alpha_{01}) > (1 + \alpha_{01})\left(\int_{+0}^{\infty} |dk_0(s)| + r_1\right), \qquad \alpha_{11} + \alpha_{10} < \infty$$
Then the trivial solution of Eq. (4.3) is globally asymptotically stable. Consider the functional
$$V(x_t) = Z^2(x_t) + (r_1 + \alpha_{00})\int_0^{\infty} |dk_1(s)| \int_{t-s}^{t} x^2(s_1)\, ds_1$$
$$+ (1 + \alpha_{01})\int_{+0}^{\infty} |dk_0(s)| \int_{t-s}^{t} x^2(s_1)\, ds_1$$
$$+ \frac{\alpha_{01} + 1}{r_1} \int_0^{\infty} |dR_1(s, \varphi)| \int_{t-s}^{t} x^2(s_1)\, ds_1 \tag{4.8}$$
Estimates (4.7) follow from the inequality
$$\int_0^{\infty} |dk_i(s)| \int_{t-s}^{t} x^2(s_1)\, ds_1 \le \alpha_{1i}\|x_t\|_B^2, \qquad i = 0, 1$$
The derivative of functional (4.8) along the trajectories of Eq. (4.3) is
$$\dot{V} \le 2x^2(t)\left[-a_0(1 - \alpha_{01}) + (1 + \alpha_{01})\left(\int_{+0}^{\infty} |dk_0(s)| + r_1\right)\right]$$
From this we arrive at the following assertion.

EXAMPLE 4.1. Consider the automatic control system represented by the structural scheme in Fig. 3.11. Let the nonlinear element satisfy the Lipschitz conditions
$$|F(x(t + \theta)) - F(y(t + \theta))| \le \int_0^{\infty} |(x(t - s) - y(t - s)|\, d_s R(s, x, y)$$
$$\int_0^{\infty} d_s R(s, x, y) \le r^2, \qquad \forall\, x, y \in Q_H$$

§4. Construction of Degenerated Functionals for Concrete Systems

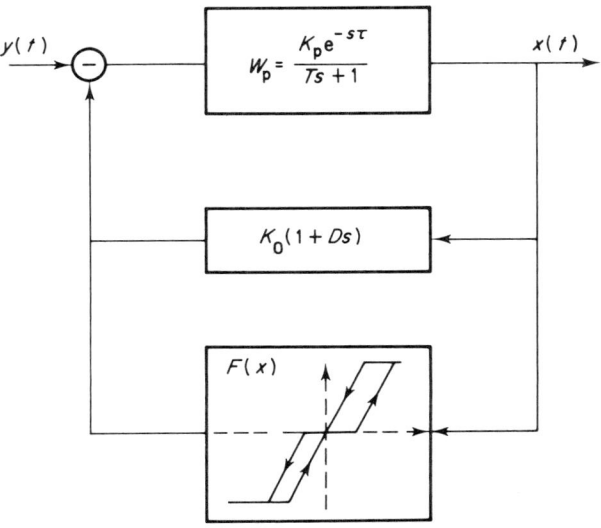

Fig. 3.11. Structural scheme of control system.

The work of the system presented in Fig. 3.11 is described by the equation

$$T\dot{x}(t) + K_0 K_p D\dot{x}(t - \tau) + x(t) + K_0 K_p x(t - \tau) \\ + K_0 F(x(t - \tau + \theta)) = K_0 y(t - \tau)$$

Designate by $z(t)$ the perturbation of the output signal corresponding to some fixed input signal $\bar{y}(t)$. The function $z(t)$ satisfies the equation

$$T\dot{z}(t) + K_0 K_p D\dot{z}(t - \tau) + z(t) + K_0 K_p z(t - \tau) \\ + K_0 [F(\bar{x}(t - \tau + \theta) + z(t - \tau + \theta)) - F\bar{x}(t - \tau + \theta))] = 0$$

Applying the statement established above, we obtain that the system, shown in Fig. 3.11 is globally asymptotically stable if

$$T < K_0 K_p D, \qquad T - K_0 K_p D > (T + K_0 K_p D) K_0 (K_p + r)$$

Depending on the concrete form of the systems under consideration, the functional $Z(\varphi)$ in Theorem 4.1 can sometimes be reasonably chosen in a different form. Let us formulate some results using this approach. Let

3. Stability of Neutral Functional Differential Equations

conditions (4.4) be satisfied, and

$$\alpha_{10} + \alpha_{01} < 1, \qquad \alpha_{20} + \alpha_{11} < \infty \tag{4.9}$$

$$\beta = \int_0^\infty dk_0(s) > r_1 \frac{1 + \alpha_{10} + \alpha_{01}}{1 - \alpha_{10} - \alpha_{01}}$$

Then the trivial solution of Eq. (4.3) is globally asymptotically stable.

Introduce the functional

$$V(x_t) = Z_1^2(x_t) + (\beta + r_1) \int_0^\infty |dk_1(s)| \int_{t-s}^t x^2(t_1)\, dt_1$$

$$+ (1 + \alpha_{10} + \alpha_{01}) r_1^{-1} \int_0^\infty dR_1(s, \varphi) \int_{t-s}^t x^2(t_1)\, dt_1$$

$$+ (\beta + r_1) \int_0^\infty |dk_0(s)| \int_{t-s}^t dt_1 \int_{t_1}^t x^2(t_2)\, dt_2 \tag{4.10}$$

Here $Z_1(\varphi)$ is determined by the formula

$$Z_1(\varphi) = \varphi(0) - \int_0^\infty \varphi(-s)\, dk_1(s) - \int_0^\infty dk_0(s) \int_{-s}^0 \varphi(t_1)\, dt_1$$

From (4.9) follows the f stability of the solution of the inequality $|Z_1(\varphi)| \le C_0$ and the estimates

$$Z_1^2(x_t) \le V(x_t) \le C_1 \|x_t\|_B^2$$

Also, it is easy to evaluate that for almost all $t > 0$

$$\dot{V} \le 2x^2(t)[-\beta(1 - \alpha_{01} - \alpha_{10}) + r_1(1 + \alpha_{01} + \alpha_{10})]$$

From this follows the validity of our statement.

Modifying functional (4.10), one can obtain different conditions of stability of Eq. (4.3). Consider some of them. Use the fact that an arbitrary function $k_0(s)$ with bounded variation can be represented in the form of the difference of two limited functions which do not decrease: $k_0(s) = k_3(s) - k_4(s)$. In this case it is sufficient to add to functional (4.10), in which $k_0(s)$ is replaced by $k_3(s)$, β by α_{03} and α_{10} by α_{13}, the expression

$$(1 + \alpha_{01} + \alpha_{13}) \int_0^\infty dk_4(s) \int_{t-s}^t x^2(t_1)\, dt_1$$

$$+ \alpha_{04} \int_0^\infty dk_1(s) \int_{t-s}^t x^2(t_1)\, dt_1$$

$$+ \alpha_{04} \int_0^\infty dk_3(s) \int_{t-s}^t dt_1 \int_{t_1}^t x^2(t_2)\, dt_2$$

§4. Construction of Degenerated Functionals for Concrete Systems

With the help of this new functional one can establish the following statement. Let conditions (4.4) be fulfilled and

$$\alpha_{01} + \alpha_{13} < 1, \qquad \alpha_{11} + \alpha_{14} + \alpha_{23} < \infty$$

$$\alpha_{03}(1 - \alpha_{01} - \alpha_{13}) > (\alpha_{04} + r_1)(1 + \alpha_{01} + \alpha_{13})$$

Then the trivial solution of Eq. (4.3) is globally asymptotically stable.

By slightly changing functionals (4.8), (4.10) one can obtain the stability conditions for the trivial solutions of the equations

$$\dot{x}(t) = -\int_0^\infty x(t-s) d_s k_0(t,s) + \int_0^\infty \dot{x}(t-s) d_s k_1(t,s) + a(t, x_t) \quad (4.11)$$

Since in this case we have to deal with rather cumbersome functionals, we shall restrict ourselves to the consideration of the simplest case, from which it is, however, easily seen what changes should be made in functionals (4.8), (4.10) for general Eqs. (4.11). Find the stability conditions of the trivial solution of the equation

$$\dot{x}(t) = -b(t)x(t-h) + c\dot{x}(t-h), \qquad t \geq 0 \quad (4.12)$$

The function $b(t) \geq 0$ is assumed to be continuous. With the help of the functional

$$V(x_t) = [x(t) - cx(t-h) - \int_{t-h}^t b(s+h)x(s)\,ds]^2$$

$$+ |c| \int_{t-h}^t b(s+2h)x^2(s)\,ds + \int_{t-h}^t b(t_1+2h)\,dt_1$$

$$\cdot \int_{t_1}^t b(t_2+h)x^2(t_2)\,dt_2$$

it is easily found that the trivial solution of Eq. (4.12) is asymptotically stable under the assumptions

$$\sup_{t \geq 0} \left\{ |c| + \int_t^{t+h} b(s)\,ds \right\} < 1$$

$$\sup_{t \geq 0} \left\{ -2b(t+h) + |c|(b(t+h) + b(t+2h)) \right.$$

$$\left. + b(t+h) \int_{t-h}^t (b(s+h) + b(s+2h))\,ds \right\} < 0$$

4.3. Use of Degenerate Functionals for Stability Investigations of RFDEs

The degenerated Lipaunov functionals may also be used with success for RFDEs. Consider the equation

$$\dot{x}(t) = -\int_0^\infty x(t-s)\,dk_0(s) + a(t, x_t), \qquad t \geq 0 \qquad (4.13)$$

studied in Chapter 2, §3.6. We shall use the assumptions and notations from Chapter 2, §6.

Let the kernel $k_0(s)$ not decrease monotonically (i.e., $\alpha_{00} = \beta_{00}$) and also

$$\beta_{10} < 1, \qquad \alpha_{00}(1 - \beta_{10}) - (1 + \beta_{10})r = \gamma > 0$$

$$\alpha_{00} + \alpha_{20} + \int_0^\infty s\,dR(s) < \infty$$

Then the trivial solution of Eq. (4.13) is globally asymptotically stable. Consider only the case $r > 0$ with the aid of the functional

$$V(t, x_t) = \left[x(t) - \int_0^\infty dk_0(s) \int_{t-s}^t x(t_1)\,dt_1\right]^2$$

$$+ (\alpha_{00} + r) \int_0^\infty dk_0(s) \int_{t-s}^t dt_1 \int_{t_1}^t x^2(t_2)\,dt_2$$

$$+ (1 + \beta_{10})r^{-1} \int_0^\infty dR(s) \int_{t-s}^t x^2(t_1)\,dt_1 \qquad (4.14)$$

Under our assumptions, we get

$$V(t, x_t) \leq C\|x_t\|_B^2$$

$$\dot{V} = 2\left[x(t) - \int_0^\infty dk_0(s) \int_{t-s}^t x(t_1)\,dt_1\right](a(t, x_t) - x(t)\alpha_{00})$$

$$+ x^2(t)[\beta_{10}(\alpha_{00} + r) + r(1 + \beta_{10})]$$

$$- (\alpha_{10} + r) \int_0^\infty dk_0(s) \int_{t-s}^t x^2(t_1)\,dt_1$$

$$- (1 + \beta_{10})r^{-1} \int_0^\infty x^2(t-s)\,dR(s)$$

§4. Construction of Degenerated Functionals for Concrete Systems

Using the inequalities

$$2x(t)a(t, x_t) \leq rx^2(t) + r^{-1} \int_0^\infty x^2(t-s)\, dR(s)$$

$$2a(t, x_t) \int_0^\infty dk_0(s) \int_{t-s}^t x(t_1)\, dt_1 \leq \beta_{10} r^{-1} \int_0^\infty x^2(t-s)\, dR(s)$$

$$+ r \int_0^\infty dk_0(s) \int_{t-s}^t x^2(t_1)\, dt_1$$

we obtain that $\dot{V} \leq -\gamma x^2(t)$. From the assumption $\beta_{10} < 1$ and Lemma 3.1 follows that f stability of the trivial solution of the inequality

$$\left| x(t) - \int_0^\infty dk_0(s) \int_{t-s}^t x(t_1)\, dt_1 \right| \leq C_1$$

Thus, from Theorem 3.1 follows the stability of the trivial solution of Eq. (4.13). Some supplementary considerations show that it is asymptotically stable [108(5)].

If the kernel $k_0(s)$ is not monotone then it may be represented as a difference of two monotonically nondecreasing bounded functions, $k_0(s) = k_1(s) - k_2(s)$. In this case it is sufficient to add the expression

$$\alpha_{12} \int_0^\infty dk_1(s) \int_{t-s}^t dt_1 \int_{t_1}^t x^2(t_2)\, dt_2$$

$$+ (1 + \beta_{11}) \int_0^\infty dk_2(s) \int_{t-s}^t x^2(t_1)\, dt_1$$

to functional (4.14) in which $k_0(s)$ is replaced everywhere by $k_1(s)$. In this case the following statement holds. Let

$$\beta_{11} < 1, \qquad \alpha_{01}(1 - \beta_{11}) > (1 + \beta_{11})(r + \alpha_{02})$$

$$\alpha_{00} + \alpha_{21} + \alpha_{12} + \int_0^\infty s\, dR(s) < \infty$$

Then the trivial solution of (4.13) is globally asymptotically stable.

EXAMPLE 4.2. Consider the equations

$$\dot{x}(t) = ax(t) - bx(t-h) + f(x(t)), \qquad t \geq 0 \tag{4.15}$$
$$f(0) = 0, \qquad |f(x_1) - f(x_2)| \leq C_1 |x_1 - x_2|, \qquad a > 0, \quad b > 0.$$

The trivial solution of (4.15) is asymptotically stable, if $bh < 1$, $b(1 - bh) > (1 + bh)(a + C_1)$. Notice that for $a > C_1$ the trivial solution of the equation $\dot{x}(t) = ax(t) + f(x(t))$ is unstable. Thus, the time lag may have the stabilizing effect on the system.

Modifying functional (4.14) one may establish some other stability conditions. Let

$$\alpha_{10} < 1, \quad (1 - \alpha_{10})\beta_{00} - (1 + \alpha_{10})r > 0$$

$$\alpha_{20} + \alpha_{00} + \int_0^\infty s \, dR(s) < \infty$$

Then the trivial solution of (4.13) is asymptotically stable. For the proof one uses functional (4.14) in which the second integral contains $|dk_0(s)|$ instead of $dK_0(s)$.

Consider now one example of a nonautonomous linear equation from which it is seen how general equations may be treated. Consider the equation

$$\dot{x}(t) = -b(t)x(t-h), \quad h > 0 \tag{4.16}$$

Here the function $b(t)$ is continuous, bounded and also $b(t) > 0$. If

$$\beta_1 = \sup_{t \geq 0} \int_t^{t+h} b(s) \, ds < 1, \quad \inf_{t \geq 0} b(t) > 0$$

then the trivial solution of (4.16) is asymptotically stable. The proof is based on the functional

$$V(t, x_t) = \left[x(t) - \int_t^{t+h} b(s)x(s-h) \, ds \right]^2$$

$$+ \int_{t-h}^t b(t_1 + 2h) \, dt_1 \int_{t_1}^t b(s+h)x^2(s) \, ds$$

Some applications of degenerated Liapunov functions for study of adaptive systems, governed by ordinary differential equations, are given by Kolmanovskii and Nosov [108(10)].

§5. INSTABILITY OF NEUTRAL-TYPE EQUATIONS

5.1. Statement of the Problem

In this section various problems of instability of NFDEs are studied. The dependence of instability conditions on the set of allowed initial disturbances is marked. We formulate general instability theorems with the use of degenerate functionals defined on the paths of disturbed motion. It is possible to extend the classical theorem of Chetaev [39] on the systems with aftereffect and to find the self-excitation conditions for certain distributed self-oscilla-

§5. Instability of Neutral-Type Equations

tory systems. For certain equations some instability conditions are obtained. As in Chapter 2, §1 we denote M a metric space of the continuous function $\varphi(t)$, $\varphi: (-\infty, 0] \to R_n$ with a metric ρ and Q_H is a sphere in the space M. Let F and G be two continuous mappings and also

$$|G(t, \varphi)| + |F(t, \varphi)| \leq F_0, \qquad t \in [0, \infty), \quad \varphi(\theta) \in Q_H \tag{5.1}$$

Consider the initial-value problem

$$(d/dt)[x(t) - G(t, x_t)] = F(t, x_t), \qquad t \geq 0 \tag{5.2}$$

$$x_0(\theta) = \varphi(\theta), \qquad -\infty < \theta \leq 0$$

Below it is assumed that a solution of problem (5.2) exists and that it is unique and continuously depends on the initial data. Let $G(t, 0) = F(t, 0) = 0$. Then Eq. (5.2) has a trivial solution.

Definition 5.1. *The trivial solution of* (5.2) *is unstable if for any positive ε and δ there exist an initial function $\varphi(\theta) \in M$ and an instant $t_1 = t_1(\delta) > 0$ such that $\rho(\varphi, 0) < \delta$ but $|x(t_1, \varphi)| \geq \varepsilon$.*

5.2. Influence of the Choice of an Admissible Class of Disturbances on Stability

A correct formulation of the stability problem for FDEs must also include the determination of the allowed initial disturbances for the problem under consideration. However, such an analysis depends essentially on the form of the problem studied and, in general, it is hardly realizable. In the theory of RFDE stability, the class of allowed disturbances is usually taken in the form of continuous functions. In this connection we must bear in mind that a solution which is stable under such disturbances will be stable also in cases when actually only a narrower class of initial disturbances is possible. On the other hand, a solution that is unstable under any continuous disturbances may prove to be stable under actually existing disturbances. This shows that the instability criteria are relatively inferior.

We now present an appropriate example. Consider an RFDE which is *stable under disturbances actually feasible in such a system and unstable in the case of arbitrary continuous disturbances.* In Shimbell [206] the following model was used to describe the behavior of the central nervous system during learning

$$\dot{x}(t) = k[x(t) - x(t-1)][N - x(t)], \qquad t > 1 \tag{5.3}$$

$$\dot{x}(t) = kx(t)[N - x(t)]$$

$$0 \leq t \leq 1, \quad x(0) = x_0, \quad 0 < x_0 < N, \quad k > 0 \tag{5.4}$$

Examine the stability of a solution $x(t) = N$ of Eq. (5.3) [108(8)]. As the class of admissible disturbances of the initial function in the interval [0, 1] let us take a set of solutions of Eq. (5.4) which have the form

$$x(t) = Nx_0[x_0 + (N - x_0)\exp(-kNt)]^{-1} \quad (5.5)$$

From (5.5) it follows that any solution $x(t)$ of Eq. (5.4) is increasing for $t \in [0, 1]$. Any solution of Eq. (5.3) which corresponds to the initial function (5.5) in [0, 1] is not decreasing in $[0, \infty)$, in addition remaining smaller or equal to N. Actually, the solution $x(t)$ cannot cross the line $x = N$, since in this case $\dot{x}(t_1) = 0$ at the point t_1, where $x(t_1) = N$; and for $t \geq t_1$ we should have $x(t) = N$. Further, by virtue of (5.3) we have $\dot{x}(1) > 0$, since $x(1) > x(0)$ and $N > x(1)$. This means that the solution $x(t)$ increases to the right of the point $t = 1$. Let t_1 be the first instant such that $\dot{x}(t_1) = 0$. Then either $x(t_1) - x(t_1 - 1) = 0$ or $N - x(t_1) = 0$. In the first case, because of Rolle's theorem there exists a point $t_2 < t_1$ such that $\dot{x}(t_2) = 0$. But this contradicts the assumption that t_1 is the first moment at which $\dot{x}(t) = 0$. In the second case, $x(t) = N$ for $t > t_1$. Hence the solution $x(t) = N$ of Eq. (5.3) is stable under arbitrary initial disturbances (5.5).

We can show, however, that the solution $x(t) = N$ will be unstable if the allowed initial disturbances in [0, 1] are taken in the form of space $C[0, 1]$. Take a twice continuously differentiable function $\varphi(\theta)$ such that $\|\varphi(\theta) - N\| \leq \delta$, $\varphi(1) < N$, $\dot{\varphi}(\theta) < 0$, $\ddot{\varphi}(\theta) < 0$, $0 \leq \theta \leq 1$. The derivatives $\dot{x}(1)$ and $\ddot{x}(1)$ are negative on the basis of (5.3). Hence the solution $x(t, 1, \varphi)$ decreases to the right of the point $t = 1$. Show that $x(t, 1, \varphi) \to -\infty$ for $t \to \infty$. Assume the contrary. Then there must exist a first point $t_1 > 1$ such that $\ddot{x}(t_1) = 0$. But $\ddot{x}(t) < 0$ for $t \in [1, t_1)$. Therefore $N - x(t_1) > 0$, $-\dot{x}(t_1) > 0$, $x(t) - x(t_1 - 1) < 0$, $\dot{x}(t_1) - \dot{x}(t_1 - 1) < 0$. Hence $\ddot{x}(t_1) = k[\dot{x}(t_1) - \dot{x}(t_1 - 1)][N - x(t_1)] + k[x(t_1) - x(t_1 - 1)][-\dot{x}(t_1)] < 0$. The contradiction that results shows that $\ddot{x}(t, 1, \varphi) < 0$ for $1 \leq t < \infty$. In addition, it is obvious that $x(t, 1, \varphi) \to -\infty$ for $t \to \infty$. Thus the solution $x(t) = N$ of (5.3) will be unstable if allowed initial disturbances coincide with the space $C[-h, 0]$.

5.3. Instability Conditions

Consider a continuous functional $V(t, x_t, Z(t, x_t))$, where $Z(t, x_t) = x(t) - G(t, x_t)$. The region $V > 0$ is a connected open region in the product $[0, \infty) \times Q_H$ which is bounded by the surface $V = 0$ and in which the functional V takes only positive values.

Theorem 5.1. *Let there exist a continuous functional $V(t) = V(t, x_t, Z(t, x_t))$ such that for $t = 0$ the region $V > 0$ has an open section whose boundary*

§5. Instability of Neutral-Type Equations

contains the element $\varphi(\theta) = 0$. Further, in the region $V > 0$ let the following conditions take place for Eqs. (5.3), (5.4), (5.5), respectively,

$$V(t) \leq q(H)$$

$$\dot{V}(t) \equiv \lim_{\Delta t \to +0} \inf \frac{1}{\Delta t} [V(t+t) - V(t)] \geq 0$$

if $V(t) \geq \alpha > 0$ *then* $\dot{V}(t) \geq \beta(\alpha) > 0$

Then the trivial solution of (5.2) is unstable.

Proof. Let $\delta > 0$ be arbitrarily small. There exists an initial function $\varphi(\theta)$ such that

$$\rho(\varphi(\theta), 0) \leq \delta, \qquad V(0) = V(0, \varphi, Z(0, \varphi)) = \alpha > 0 \qquad (5.6)$$

Show that a solution $x(t, \varphi)$ of (5.2) with such an initial function leaves the sphere Q_H for $t \to \infty$. The solution $x(t, \varphi)$ cannot leave the region $V > 0$ by passing through the part of the boundary on which $V = 0$. In fact, if $t_1 > 0$ is such that

$$V(t_1) = 0, \qquad V(t) > 0 \quad \text{for} \quad 0 \leq t < t_1 \qquad (5.7)$$

then by virtue of (5.4) it follows that $\dot{V}(t) \geq 0$, $0 \leq t < t_1$. Hence $V(t) \geq V(0) = \alpha > 0$, and by continuity $V(t_1) \geq V(0) = \alpha > 0$, which contradicts (5.7). Hence the solution $x(t, \varphi)$ cannot leave the region $V > 0$ across the part of the boundary on which $V = 0$. But the solution $x(t, \varphi)$ cannot always remain in the region $V > 0$. Assuming this, we would successively obtain

$$\dot{V}(t) \geq 0, \qquad V(t) \geq V(0) = \alpha > 0, \qquad \dot{V}(t) \geq \beta(\alpha) > 0 \qquad (5.8)$$

We find by integrating (5.8) that

$$V(t, x_t, Z(t, x_t)) \geq V(0, \varphi, Z(0, \varphi)) + t\beta(\alpha) \qquad (5.9)$$

If $t > \beta^{-1}(\alpha)[q(H) - \alpha]$, then inequality (5.9) contradicts condition (5.3). So the solution $x(t, \varphi)$ must leave the region $V > 0$ through the part of the boundary on which $\rho(x_t, 0) = H$. This completes the proof of Theorem 5.1.

REMARK 5.1. Suppose that we can select a subregion D which belongs to the region $V > 0$ and such that all the conditions of Theorem 3.1 are fulfilled in D. Suppose also that $x(t, \varphi) \in D$ for all $t > 0$. Then the assertion of Theorem 3.1 is valid. A region D which satisfies the preceding conditions is called a *sector* [191].

EXAMPLE 5.1. Consider the equation

$$\dot{x}(t) - c\dot{x}(t - \tau) = a(t)x(t) + b(t) \times (t - \tau) \quad (5.10)$$

$$c \geq 0, \quad a(t) \geq 0, \quad b(t) \geq 0, \quad a(t) + b(t) \geq A > 0, \quad \tau > 0 \quad (5.11)$$

Introduce the functional $V(t) = Z(t, x_t) = x(t) - cx(t - \tau)$. Choose the initial function $\varphi(\theta)$ such that $\varphi(\theta) > 0$, $\|\varphi(\theta)\| \leq \delta$, $Z(0, \varphi) = \varphi(\theta) - c\varphi(-\tau) = \alpha > 0$. Show that $x(t, \varphi) > 0$ for $t > 0$. Assume the contrary and let $t_1 > 0$ be the first instant such that $x(t_1, \varphi) = 0$. For $0 < t < t_1$,

$$\dot{Z}(t, x_t = \dot{x}(t) - c\dot{x}(t - \tau) = a(t)x(t) + b(t)x(t - \tau) > 0$$

Hence $Z(t, x_t) \geq Z(0, \varphi) = \alpha > 0$. By going to the limit for $t \to t_1 - 0$, we obtain $Z(t_1, x_{t_1}) \geq \alpha > 0$. Therefore $x(t_1) = Z(t_1, x_{t_1}) + cx(t_1 - \tau) \geq \alpha > 0$. The latter inequality contradicts the preceding assumption. Thus for any $t > 0$ we get $\dot{V}(t) = a(t)x(t) + b(t)x(t - \tau) \geq A\alpha > 0$.

By virtue of Remark 5.1 the trivial solution of (5.10) will be unstable under conditions (5.11). In the same way it is possible to prove that the trivial solution of the equation $\dot{x}(t) - c\dot{x}(t - \tau) = g(x_t)$, $g(0) = 0$, $c > 0$ is unstable if the functional $g(x_t(\theta)) > 0$ for $\min_\theta x_t(\theta) > 0$, $-\tau \leq \theta \leq 0$.

5.4. Connection between Instability of NFDEs and Ordinary Differential Equations

As a consequence of Theorem 5.1, investigate the connection between the instability of an ordinary differential equation

$$\dot{x}(t) = F(t, x_t), \quad x(t_0) = x_0, \quad t_0 \geq 0, \quad F(t, 0) = 0 \quad (5.12)$$

and an NFDE

$$\dot{Z}(t, x_t) = F(t, Z(t, x_t)), \quad x_0(\theta) = \varphi(\theta)$$
$$Z(t, x_t) = x(t) - G(t, x_t), \quad G(t, 0) = 0 \quad (5.13)$$

Theorem 5.2. *If the trivial solution of ordinary equation (5.12) is unstable then the trivial solution of NFDE (5.13) is also unstable.*

Proof. By the inversion theorem for Chetaev's theorem [114(5)] there exists for (5.12) a function $W(t, x)$ which satisfies the conditions of Chetaev's theorem. In particular, by virtue of (5.12)

$$\dot{W}(t, x) = \frac{\partial W}{\partial t} + \sum_{i=1}^{n} \frac{\partial W}{\partial x_i} \dot{x}_i = \Phi(t, x) \quad (5.14)$$

Now consider the functional $W(t, Z(t, x_t))$. Using (5.14) and (5.13) we find that $W(t, Z(t, x_t)) = \Phi(t, x_t))$. The validity of the other assumptions of

§5. Instability of Neutral-Type Equations 147

Theorem 5.1 about the functional $W(t, Z(t, x_t))$ is obvious. Hence the functional $W(t, Z(t, x_t))$ satisfies all the conditions of Theorem 5.1. So the trivial solution of (5.13) is unstable.

EXAMPLE 5.2. Consider the equation

$$(d^n/dt^n)[x(t) + cx(t - \tau)] = f(x(t) + cx(t - \tau)) \qquad (5.15)$$

Let c be a constant, $n \geq 3$, and let $y = 0$ be an isolated root of the equation $f(y) = 0$. Then the trivial solution of the equation $y^n(t) = f(y)$ is unstable ([191], p. 148). According to Theorem 5.2, the trivial solution of (5.15) is also unstable.

5.5. Other Instability Conditions

Denote $\omega_i(u)$ as scalar nondecreasing functions of the argument $u \geq 0$, such that $\omega_i(0) = 0$ and $\omega_i(u) > 0$ for $u > 0$.

Theorem 5.3. *Let all the conditions of Theorem 5.1 hold except conditions (5.3) and (5.5), which are replaced in the region $V > 0$ by*

$$V(t, x_t, Z(t, x_t)) \geq \omega_2(|Z(t, x_t)|) \qquad (5.16)$$

$$\dot{V}(t, x_t, Z(t, x_t)) \geq \omega_3(|Z(t, x_t)|) \qquad (5.17)$$

Then the trivial solution of (5.2) is unstable.

Proof. Choose any initial function φ satisfying the conditions (5.6). By an argument similar to that made in the proof of Theorem 5.1, it is easy to verify that for this initial function the solution $x(t, \varphi)$ cannot leave the region across that part of the boundary at which $V = 0$. However, the solution $x(t, \varphi)$ cannot remain all the time in the region $V > 0$. Otherwise

$$\dot{V}(t) \geq 0, \qquad V(t) \geq V(0) = \alpha > 0$$

Take $\beta > 0$ such that $\omega_2(\beta) = \alpha$. Then $\omega_2(\beta) = \alpha = V(0) \leq V(t) \leq \omega_2(|Z(t, x_t)|)$. Hence $\beta \leq |Z(t, x_t)|$. Consequently,

$$\dot{V}(t) \geq \omega_3(\beta) > 0 \qquad (5.18)$$

Integrating (5.18) we obtain a contradiction with the boundedness of the functional $V(t, x_t, Z(t, x_t))$ in the sphere Q_H: $V(t, x_t, Z(t, x_t)) \leq \omega_2(|Z(t, x_t)|) \leq \omega_2(H + F_0)$. Theorem 5.3 is proved.

EXAMPLE 5.3. Given the equation

$$(d/dt)[x(t) - g(t, x(t - h))] = ax(t), \qquad a > 0, \quad h > 0$$

$$g: R_1 \times R_1 \to R_1 \qquad (5.19)$$

Let
$$V(t, x_t, Z(t, x_t)) = Z^2(t, x_t) - a \int_{t-h}^{t} x^2(s)\, ds \tag{5.20}$$

$$Z(t, x_t) = x(t) - g(t, x(t-h))$$

Functional (5.20) satisfies condition (5.16). Further if $g^2(t, x(t-h)) \le x^2(t-h)$, then

$$\begin{aligned}\dot V &= 2ax(t)Z(t,x_t) - ax^2(t) + ax^2(t-h)\\ &= ax^2(t) - 2ax(t)g(t,x(t-h)) + ag^2(t, x(t-h))\\ &\quad + ag^2(t, x(t-h)) + ax^2(t-h) \ge aZ^2(t, x_t)\end{aligned}$$

Hence by virtue of Theorem 3.3 the trivial solution of (5.19) is unstable.

5.6. Instability of Equations with Bounded Delay

Investigate Eqs. (5.2) with finite or infinite delay but such that the initial set coincides with the interval $[-h, 0]$. So F and G are continuous functionals mapping from $[0, \infty) \times C[-h, t]$ into R_n.

Theorem 5.4. *Let there exist a continuous functional $V(t, x_t, Z(t, x_t))$ and the region $V > 0$ satisfy the conditions of Theorem 5.1. Let in the region $V > 0$*

$$V(t, x_t, Z(t, x_t)) \le \omega_4(\|x_t\|) \tag{5.21}$$

$$|G(t+s, x_{t+s}) - G(t, x_t)| \le \omega_5(s) + v \max_{-h \le \tau \le t} |x(\tau+s) - x(\tau)| \tag{5.22}$$

$$0 < v < 1, \qquad s \ge 0, \qquad \forall\, x_t \in Q_H, \quad t \ge 0 \tag{5.23}$$

$$\dot V(t, x_t, Z(t, x_t)) \ge \omega_6(|x(t)|)$$

Then the trivial solution of (5.2) is unstable.

Proof. The solution $x(t, \varphi)$ cannot leave the region $V > 0$ through the part of the boundary $V = 0$. Show that the solution $x(t, \varphi)$ cannot remain in the region $V > 0$. Really, let the initial function satisfy (5.6) and $\mu > 0$ such that $\omega_4(\mu) = \alpha$. Then $\omega_4(\mu) = \alpha = V(0) \le V(t) \le \omega_4(\|x_t\|)$. Consequently, there exists a sequence $\{t_i\}$ such that $t_i \to \infty$ and $|x(t_i)| = \mu$. Show that the function $x(t) = x(t, \varphi)$ is uniformly continuous in $[-h, \infty)$. By using (5.1), (5.2) and (5.22) for any $\Delta > 0$, we get

$$|x(t+\Delta) - x(t)| = |G(t+\Delta, x_{t+\Delta}) - G(t, x_t) + \int_{t}^{t+\Delta} F(s, x_s)\, ds| \tag{5.24}$$

Denote $\rho(\eta)$ the function

$$\rho(\eta) = \sup_{t \ge 0} \sup_{\Delta \le \eta} |x(t+\Delta) - x(t)|$$

§5. Instability of Neutral-Type Equations

From (5.24) and (5.22) it follows that
$$(1 - v)\rho(\eta) \leq \max_{\Delta \leq \eta} \max_{-h \leq \theta \leq 0} |x(\theta + \Delta) - \varphi(\theta)| + \eta F_0 + \omega_5(\eta)$$

The right-hand side of this inequality tends to zero for $\eta \to 0$ due to the uniform continuity of $x(t, \varphi)$, $t \in [-h, \eta]$. Hence the uniform continuity of $x(t, \varphi)$, $t \in [-h, \infty)$ is proved. Therefore there exists an η such that $|x(s)| \geq \mu/2$ for $t_i - \eta \leq s \leq t_i + \eta$ and any i. Integrating (5.23) we obtain for $t \to \infty$

$$V \geq \alpha + \int_0^t \omega_6(|x(s)|) \, ds \geq \alpha + 2N(t)\omega_6(\mu/2)$$

where $N(t)$ is a number of points t_i such that $t_i \in [0, t]$. Theorem 5.4 is proved.

EXAMPLE 5.4. Obtain the instability conditions for a scalar NFDE

$$\dot{x}(t) = a_0 x(t) + \int_0^h x(t - s) \, dk_0(s) + \int_0^h \dot{x}(t - s) \, dk_1(s) + a(t, x_t)$$
$$k_1(s) = \int_0^s \lambda(t) \, dt + \sum_{h_n \leq t} \mu_n, \qquad h_n \geq 0$$
(5.25)

Here the continuous functional $a(t, \varphi)$ is defined on $[0, \infty) \times \text{CB}[-\infty, 0]$ and also

$$a(t, 0) = 0, \qquad |a(t, \varphi)|^2 \leq \int_0^h |\varphi(-s)|^2 \, dR(s)$$

Kernel $R(s)$ is nondecreasing and bounded. Introduce a functional V,

$$V(x_t, Z(t, x_t)) = Z^2(t, x_t) - (r + \alpha_{00}) \int_0^h dk_1(s) \int_{t-s}^t x^2(u) \, du$$
$$- (1 + \alpha_{01}) \int_0^h |dk_0(s)| \int_{t-s}^t x^2(u) \, du$$
$$+ (1 + \alpha_{01})r^{-1} \int_0^h dR(s) \int_{t-s}^t x^2(u) \, du$$

Here

$$r^2 = \int_0^h dR(s), \qquad \alpha_{0i} = \int_0^h |dk_i(s)|$$

$$Z(t, x_t) = x(t) - \int_0^h x(t - s) \, dk_1(s)$$

Relations (5.25), yield $\dot{V}(t) \geq 2x^2(t)[a_0(1 - \alpha_{01}) - (1 + \alpha_{01})(r + \alpha_{00})]$. By virtue of Theorem 5.4 the trivial solution of (5.25) is unstable under condition

$a_0(1 - \alpha_{01}) > (1 + \alpha_{01})(r + \alpha_{00})$. By using the methods of functionals V construction described earlier in this book, it is possible to obtain other conditions of instability of system (5.25).

5.7. Distributed Self-Oscillatory Systems

Consider the distributed lossless system from Chapter 1, §2 described by the equation

$$\frac{1}{C_1} \frac{d}{dt} [x(t) - Kx(t - \tau)]$$
$$= \alpha - \frac{1}{Z} x(t) - \frac{K}{Z} x(t - \tau) - g(x(t)) + Kg(x(t - \tau)), \qquad \tau = \frac{2l}{s} \quad (5.26)$$

Assume that the nonlinear element $g(V)$ has a negative differential resistance at certain points. Such an element can be the tunnel diode whose current–voltage curve is plotted in Fig. 3.12 [147]. Find the conditions of self-excitation of oscillations which are equivalent to the conditions of instability of the solution $x(t) = u_0$. The value u_0 satisfies the equation $0 = \alpha - Z^{-1}u_0 - KZ^{-1}u_0 - g(u_0) + Kg(u_0)$. Denote $y(t) = x(t) - u_0$. Equation (5.26) in the neighborhood of u_0 can be written in the form

$$\frac{1}{C_1} \frac{d}{dt} [y(t) - Ky(t - \tau)]$$
$$= (S - Z^{-1})y(t) - K(S + Z^{-1}) + 0(y(t)) + 0(y(t - \tau)) \quad (5.27)$$

where S is the steepness of the current–voltage curve at the point u_0; i.e., $s = -\dot{g}(u_0)$. Show that the *condition of self-excitation* of the distributed

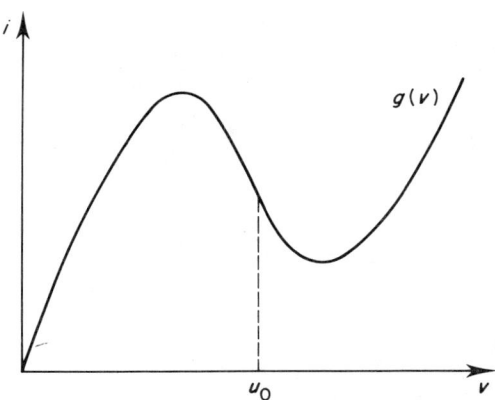

Fig. 3.12. Current-voltage curve.

self-oscillatory system represented in Fig. 1.12 *has the form* $S > R^{-1}$. Study for this purpose the directly nonlinear equation (5.27). Take an $\varepsilon > 0$ so small that

$$S > (1 + K + 2\varepsilon)(1 - K)^{-1}Z^{-1} \quad (5.28)$$

and also in the sphere Q

$$\frac{1}{C_1}\frac{d}{dt}[y(t) - Ky(t - \tau)]$$
$$\geq (S - Z^{-1} - \varepsilon)y(t) - [(K(S + Z)^{-1}) - \varepsilon]y(t - \tau) \quad (5.29)$$

Consider the functionals

$$V = y(t) - Ky(t - \tau) - [K(S + Z)^{-1} + \varepsilon]\int_{t-\tau}^{t} y(s)\,ds \quad (5.30)$$

The region $B = (V > 0) \cap (y > 0) \cap Q_\varepsilon$ is a sector. The trajectories of system (5.26) cannot cross the boundary $V = 0$ because

$$\dot{V} = [S - Z^{-1} + K(S + Z^{-1}) + 2\varepsilon]y(t) \geq 0.$$

In a manner similar to Example 5.1, we conclude that $y(t) > 0$ for $t > 0$. Otherwise, if $y(t_1) = 0$, then

$$V(t) \geq V(0) \geq \alpha, \; y(t_1) = V(t_1) + Ky(t_1 - \tau)$$
$$+ [(K(S + Z)^{-1} + \varepsilon]\int_{t_1-\varepsilon}^{t_1} y(s)\,ds.$$

Hence the region B is a sector in which $\dot{V} \geq 0$. By virtue of Remark 5.1, the solution $x(t) = u_0$ is unstable.

§6. ASYMPTOTIC PROPERTIES OF NEUTRAL-TYPE EQUATIONS

6.1. Some Introductory Remarks

One of the important problems in the qualitative theory of NFDEs is the study of the asymptotic behaviour of its solutions. Asymptotic behavior for $t \to \infty$ of NFDE solutions may be different. They can tend to stationary or periodic solutions or their behavior can be more intricate. Neutral functional differential equations have properties that are similar to the properties of difference equations for some cases and properties that are similar to the properties of differential equations for other cases. This duality of NFDEs is essential for a study of the asymptotic behavior of their solutions. In this section some results of both types are described. The method of investigation that we use is based essentially on degenerate Liapunov functionals. It is possible with the aid of degenerate functionals to separate the cases when the asymptotic behaviour of NFDEs is similar to that of difference equations from the cases in which this is not valid.

6.2. Basic Definitions

Consider the NFDE [108(7)]

$$\dot{Z}(t) = F(t, Z), \qquad Z(t) = x(t) - G(t, x_t), \qquad t \geq 0 \qquad (6.1)$$

Here F and G are given continuous mappings, $F(t, x)$, $F: [0, \infty) \times R_n \to R_n$; $G(t, x_t)$, $G: [0, \infty \times C[-h, 0] \to R_n$. The solution of (6.1) is determined by the initial data

$$x_0 = \varphi, \qquad \varphi \in C[-h, 0] \qquad (6.2)$$

Let $|F(t, x)| \leq M_H$ for $|x| \leq H$ and $|G(t, \varphi) - G(t, \psi)| \leq K_H \|\varphi - \psi\|$ for $\|\varphi\| \leq H$, $\|\psi\| \leq H$. Assume that there exists a unique solution of problem (6.1), (6.2). Scalar equations (6.1) named completely integrable are considered in Sharkovskii and Romanenko [200]. Let $M \subset C[-h, 0]$ be a closed set. The distance ρ between the element $\varphi \in C[-h, 0]$ and the set M is equal to $\rho(\varphi, M) = \min_\Psi \|\varphi - \Psi\|$, $\Psi \in M$. A set $M \subset C[-h, 0]$ is called an *invariant set* for Eq. (6.1) if for arbitrary $\varphi \in M$, the solution $x_t(\varphi) \in M$.

Definition 6.1. *The closed invariant set M of Eq. (6.1) is called stable if for any $\varepsilon > 0$ there exists $\delta > 0$ such that $\rho(x_t(\varphi), M) \leq \varepsilon, t > 0$, for $\varphi: \rho(\varphi, M) < \delta$. If, in addition, $\lim \rho(x_t(\varphi), M) = 0$, $t \to \infty$, $\forall \varphi \in \Omega$, then the set M is called asymptotically stable and the set Ω is called the attraction domain of the set M.*

Stability of invariant sets is investigated with the aid of the functional inequality

$$|Z(t, y_t)| = |y(t) - G(t, y_t)| \leq f(t), \qquad f \in R_1 \qquad (6.3)$$

$$y_0(\varphi) = \varphi, \qquad \varphi \in C[-h, 0] \qquad (6.4)$$

Let $M_1 \subset C[-h, 0]$ be a closed invariant set of equation

$$Z(t, y_t) = y(t) - G(t, y_t) = 0 \qquad (6.5)$$

Definition 6.2. *The closed invariant set M_1 of functional Eq. (6.5) is called f stable if for any $\varepsilon > 0$ there exists $\delta > 0$ such that for all solutions of inequality (6.3), (6.4) we have $\rho(y_t, M_1) \leq \varepsilon$ if $\rho(\varphi, M_1) \leq \delta$ and $f(t) \leq \delta$. The f stable set*

§6. Asymptotic Properties of Neutral-Type Equations

M_1 is called asymptotically f stable if $\lim \rho(y_t(\varphi), M_1) = 0$, $t \to \infty$, for all $\varphi \in \Omega \subset C[-h, 0]$ and for all $f(t)$ such that $f(t) \to 0$, $t \to \infty$.

For the case in which the sets M and M_1 contain the only zero element, Definitions 6.1 and 6.2 coincide with Definitions 3.2.

6.3. Stability of Invariant Sets

Theorem 6.1. *Let the trivial solution of the ordinary differential equation*

$$\dot{x}(t) = F(t, x(t)) \tag{6.6}$$

be uniformly stable (uniformly asymptotically stable). Further assume that the bounded invariant set M_1 of (6.5) is f stable (asymptotically f stable). Then M_1 will be a stable (asymptotically stable) invariant set of Eq. (6.1).

Proof. By virtue of the inversion theorem [114(5)] there exists for Eq. (6.6) a Liapunov function $V(t, x)$ such that

$$\omega_1(|x|) \le V(t, x) \le \omega_2(|x|), \qquad \dot{V}_{(6.6)} \le 0, \quad [\dot{V}_{(6.6)} \le -\omega_3(|x|)]$$

Here $\omega_i(u)$ are continuous scalar nondecreasing functions such that $\omega_i(0) = 0$, $\omega_i(u) > 0$ for $u > 0$ and $\dot{V}_{(6.6)}$ is the derivative of the function V along the trajectories of (6.6). Now consider the functional $V(t, Z(t, x_t))$. Clearly, $\omega_1(|Z(t, x_t)|) \le V(t, Z(t, x_t)) \le \omega_2(|Z(t, x_t)|)$, $\dot{V}(6.1) \le 0$.

Assume any $\varepsilon > 0$. From the f stability of the set M_1 follows the existence of a $\delta > 0$ such that $\rho(y_t(\varphi), M_1) \le \varepsilon$ for $\rho(\varphi, M_1) \le \delta$ and $f(t) \le \delta$. Choose $\delta_1 > 0$ such that $\omega_2(\delta_1 + K_H \delta_1) \le \omega_1(\delta)$. Let $\rho(\varphi, M_1) = \|\varphi - \psi\| \le \delta_1$, $\|\varphi\| \le H$ and $\|\psi\| \le H$. Since $Z(t, \psi) = 0$ for $\psi \in M_1$,

$$|Z(t, \varphi)| = |Z(t, \psi + (\varphi - \psi))| \le |Z(t, \psi)|$$
$$+ |Z(t, \psi + (\varphi - \psi)) - Z(t, \psi)| \le (K_H + 1)\delta_1.$$

Using the properties of functional $V(t, Z(t, x_t))$, we have

$$\omega_1(|Z(t, x_t)|) \le V(t, Z(t, x_t)) \le V(0, Z(0, \varphi))$$
$$\le \omega_2(\delta_1 + K_H \delta_1) \le \omega_1(\delta)$$

The monotonicity of the function $\omega_1(u)$ implies that $|Z(t, x_t)| \le \delta$. But the set M_1 is f stable. Hence $\rho(X_t, M_1) \le \varepsilon$; i.e., the set M_1 is stable. Similarly, asymptotic stability of the set M_1 may be established (see Theorem 3.3). Theorem 6.1 is proved.

EXAMPLE 6.1. Consider the scalar equation

$$\dot{Z}(t) = f(Z(t)), \qquad Z(t) = x(t) - ax(t-1)[1 - x(t-1)]$$
$$1 < a < 3, \quad t \ge 0 \tag{6.7}$$

Assume that the trivial solution of the equation $\dot{x}(t) = f(x(t))$, $f(0) = 0$, is asymptotically stable. For Eq. (5.7), difference Eq. (6.5) has the form

$$x(t) = ax(t-1)[1 - x(t-1)] \qquad (6.8)$$

Equation (6.8) was studied in Sharkovskii and Romanenko [200]. For $1 < a < 3$ there exists two stationary solutions of (6.8): $x_1(t) = 0$ and $x_2(t) = 1 - a^{-1}$. Denote $u(t) = x(t) - x_2(t)$. Then by using (6.8) we get

$$|u(t) - (2-a)u(t-1) + au^2(t-1)| \le f(t)$$
$$|u(t)| \le |2-a||u(t-s)| + au^2(t-1) + f(t) \qquad (6.9)$$

Let $\varepsilon > 0$ be sufficiently small. Choose $\delta \in (0, \varepsilon)$ such that $|2-a|\varepsilon + a\varepsilon^2 + \delta \le \varepsilon$. Remark that such a δ always exists for $1 < a < 3$ and sufficiently small ε. For these ε and δ from (6.9) it follows that for φ, $\|\varphi\| \le \delta$ and f, $\|f\| \le \delta$,

$$|u(t)| \le |2-a||\varphi(t-1)| + a\varphi^2(t-1) + f(t)$$
$$\le |2-a|\delta + a\delta^2 + \delta \le \varepsilon, \qquad 0 < t < 1 \qquad (6.10)$$

By the step method similar to (6.10) we obtain $|u(t)| \le \varepsilon$, $t > 0$. Hence the set M_1 is f stable. The asymptotic stability of M_1 follows now from Kolmanovskii and Nosov [108(5)], p. 177, Lemma 2.2. The attraction domain of the set M_1 consists of functions $\varphi \in C[-h, 0]$ such that $\rho(\varphi, M_1) \le a^{-1}(1 - |2-a|)$. By virtue of Theorem 6.1 the set M_1 is asymptotically stable for NFDE (6.7).

6.4. Instability

Investigate the conditions for which Eq. (6.5) does not determine the asymptotic behaviour of the solutions of NFDE (6.1) solutions. Let the initial functions $\varphi \in C[-h, 0]$ be bounded, $\|\varphi\| \le H$. Assume that the corresponding solutions $y_t(\varphi)$ of (6.5) are defined for all $t > 0$. The set of such solutions is designated by $N(H)$.

Definition 6.3. *The set $N(H)$ is f unstable if for any $\varepsilon > 0$ there exist φ, $\|\varphi\| \le H$ and $q = q(\varepsilon)$ such that $|Z(q, x_q(\varphi)) - Z(q, y_q)| = |Z(q, x_q(\varphi))| \ge \varepsilon$.*

If the set $N(H)$ is f unstable, then in general it does not define the asymptotic properties of the solutions of Eqs. (6.1).

Theorem 6.2. *Let the trivial solution Eq. (6.6) be unstable. Then the set $N(H)$ is f unstable.*

§6. Asymptotic Properties of Neutral-Type Equations

Proof. According to the inversion theorem [114(5)], there exists for Eq. (6.6) a function $V(t, x)$ satisfying the conditions of the Chetaev theorem. Consider the functional $V(t, Z(t, x_t)) = V(t)$. There are elements φ with an arbitrarily small norm in the domain $V > 0$. In this domain the conditions $\dot{V}_{(6.1)} \geq 0$ and $\dot{V}_{(6.1)} \geq \beta > 0$ if $V(t) \geq \alpha > 0$ are fulfilled. In addition, the solution $x_t(\varphi)$ leaves the set $N(H)$. In fact, if $V(0, Z(0, \varphi)) = \alpha > 0$, then $\dot{V}_{(6.1)}(t) \geq 0$. Hence $V(t) \geq V(0) = \alpha > 0$ and $\dot{V}_{(6.1)}(t) \geq \beta > 0$, $V(t) \geq \beta t \to \infty$ for $t \to \infty$. But $V(t, Z)$ is continuous and bounded in Z for all t. Consequently, the functional Z cannot be bounded for all t. So $Z(t, x_t)$ will be greater than any arbitrary $\varepsilon > 0$ as soon as t is sufficiently large. Theorem 6.2 is proved.

EXAMPLE 6.2. Consider once more Eq. (6.7). Assume that Eq. (6.7) is unstable and $0 < a < 1$. Then there exists the unique asymptotically stable solution $y(t) = 0$ of (6.8). But for NFDE (6.7) this solution will be unstable. So in the considered case, the behaviour of solutions of difference equations (6.8) and NFDE (6.7) are quite different.

6.5. Interconnection between Functional and Differential Equations

Sometimes the asymptotic properties of solutions to NFDE (6.1) are determined both by the behaviour of solutions of functional equations (6.5) and by the behavior of differential Eq. (6.6). Consider Eq. (6.1) with

$$Z(t, x_t) = x(t) + q(t, x(t-1), x(t-2), \ldots, x(t-p))$$
$$x(t) \in R_n, \quad q: R_n \times \cdots \times R_n \to R_n \tag{6.11}$$

Here q is a continuous function and

$$|q(t, x(t-1), x(t-2), \ldots, x(t-p))|$$
$$\leq |x(t-1)| + \cdots + |x(t-p)| \tag{6.12}$$

Theorem 6.3. *Let the trivial solution of Eq. (6.6) be globally exponentially stable and condition (6.12) be fulfilled. Then all solutions of Eq. (6.1) with $Z(t, x_t)$ given by (6.11) are bounded for $t \geq 0$ and its trivial solution is stable.*

Proof. Global exponential stability of the trivial solution of Eq. (6.6) means that there exist two constants $C(x_0) > 0$ and $\alpha > 0$ such that

$$|x(t, x_0)| \leq C(x_0)e^{-\alpha t}, \quad \forall x_0 \tag{6.13}$$

So any solution of (6.1) by (6.13) satisfies the estimate

$$|Z(t, x_t(\varphi))| \leq C(Z(0, \varphi))e^{-\alpha t}$$
$$\leq C_1(\|\varphi\|)e^{-\alpha t}, \quad C_1(\|\varphi\|) \leq C(p\|\varphi\|) \tag{6.14}$$

From (6.14) and (6.11) it follows that

$$|x(t) + q(t, x(t-1), \ldots, x(t-p))| \le C_1(\|\varphi\|)e^{-\alpha t}$$

Hence

$$\begin{aligned}|x(t)| &\le |x(t-1)| + |x(t-2)| + \cdots + |x(t-p)| + C_1(\|\varphi\|)e^{-\alpha t} \\ &\le |x(t-2)| + \cdots + |x(t-p)| + |x(t-p-1)| \\ &\quad + C_1(\|\varphi\|)[e^{-\alpha t} + e^{-\alpha(t-1)}] \\ &\le |\varphi(t-N)| + \cdots + |\varphi(t-N-p)| \\ &\quad + C_1(\|\varphi\|)[e^{-\alpha t} + e^{-\alpha(t-1)} + \cdots + e^{-\alpha(t-N+1)}]\end{aligned} \qquad (6.15)$$

Here the number N is such that $N - 1 < t \le N$. Designate $\gamma = \exp(-\alpha) < 1$. Then inequality (6.15) takes the form

$$|x(t)| \le p\|\varphi\| + C_1(\|\varphi\|)[\gamma^{N-1} + \gamma^{N-2} + \cdots + 1]$$
$$\exp[-\alpha(t - N + 1)] \le p\|\varphi\| + C_1(\|\varphi\|)(1 - \gamma)^{-1} \qquad (6.16)$$

So all solution of (6.1) are bounded. But estimates (6.16) is uniform in φ with $\|\varphi\| \le H$. Therefore from (6.16) follows the stability of the trivial solution of Eq. (6.11).

The Theorem 6.3 statement is also valid if condition (6.12) is exchanged for the following one

$$\begin{aligned}|q(t, x(t-1)), x(t-2), \ldots, x(t-p)| \\ \le K[|x(t-1)| + \cdots + |x(t-p)|], \qquad K \exp(-\alpha) < 1\end{aligned} \qquad (6.17)$$

For $K > 1$ the trivial solution of the difference equation $x(t) + q(t, x(t-1), \ldots, x(t-p)) = 0$ is in general unstable in the Liapunov sense. Condition (6.17) coordinates the order of this instability with the speed of convergence to zero of all solutions of ordinary differential Eq. (6.6). In addition, all solutions of NFDE (6.1) are bounded and the trivial solution of (6.1) is stable.

§7. LIAPUNOV FUNCTIONALS DEPENDING ON DERIVATIVES

7.1. General Stability Theorem

Let us denote by W a normed space of absolutely continuous functions $\varphi(\theta)$ defined on $[-h, 0]$ with square integrable derivatives. Define the norm in W

$$\|\varphi\|_W^2 = \varphi^{2(0)} + \int_{-h}^{0} \dot\varphi^2(\theta)\, d\theta$$

§7. Liapunov Functionals Depending on Derivatives

Denote Q_H a sphere in W: $Q_H = \{\varphi(\theta)): \|\varphi(\theta)\|_W \leq H\}$. Consider the following NFDE

$$\begin{aligned} \dot{x}(t) &= f(t, x_t, \dot{x}_t), & x(t) &\in R_n, & t &\geq t_0 \\ x_{t_0}(\theta) &= \varphi(\theta), & \dot{x}_{t_0} &= \dot{\varphi}(\theta), & -h &\leq \theta \leq 0 \end{aligned} \quad (7.1)$$

Let the map $f: R_1 \times Q_H \to R_n$ be continuous and satisfy the Lipschitz condition in the second and third arguments, and let the Lipschitz constant in the third argument be less than 1. Then, as in Theorem 1.3.2, one can prove that for any $\varphi(\theta) \in W$ there exists a unique solution of NFDE (7.1) such that $x_t(\theta) \in W$ for all $t \geq t_0$. Suppose that

$$f(t, 0, 0) = 0, \quad |f(t, \varphi, \dot{\varphi})| \leq M, \quad \|\varphi\|_W \leq H, \quad t \in R \quad (7.2)$$

Definition 7.1. *The trivial solution $x(t) = 0$ of Eq. (7.1) is said to be stable if for any $\varepsilon > 0$, $t_0 \in R$, there is a $\delta = \delta(\varepsilon, t_0)$ such that $\|\varphi\|_W \leq \delta(\varepsilon, t_0)$ implies $x(t, t_0, \varphi)| \leq \varepsilon$. The solution $x(t) = 0$ is said to be asymptotically stable if it is stable and $\varphi(t) \in \Omega(t_0) \subset W$ implies $x(t, t_0, \varphi) \to 0$ as $t \to \infty$. The domain $\Omega(t_0)$ is called an attraction one at moment t_0.*

Other definitions of stability can be introduced in a manner analogous to that presented in Chapter 2, §1.

Let the functional $V(t, \varphi, \dot{\varphi})$ be continuous in the sphere Q_H and satisfy the Lipschitz condition in φ and $\dot{\varphi}$. Denote by $V(t) = V(t, x_t, \dot{x}_t)$, where $x(t)$ is a solution of Eq. (7.1). Suppose that the functional $V(t)$ is absolutely continuous in t for any solution $x(t)$ of Eq. (7.1). We denote $\dot{V}(t)$ the upper right-hand derivative of $V(t, x_t, \dot{x}_t)$ along the solution of Eq. (7.1):

$$\dot{V}(t) = \overline{\lim_{\Delta t \to +0}} [V(t + \Delta t) - V(t)]/(\Delta t)$$

Denote $\omega_i(u)$ continuous nondecreasing functions such that $\omega_i(0) = 0$, $\omega_i(u) > 0$ for $u > 0$.

Theorem 7.1. *Let the continuous functional $V(t, x_t, \dot{x}_t)$ exist and satisfy conditions*

$$\omega_1(\|x_t\|_W) \leq V(t, x_t, \dot{x}_t) \leq \omega_2(\|x_t\|_W) \quad (7.3)$$

$$\dot{V}(t) \leq 0 \quad (7.4)$$

Then the trivial solution of Eq. (7.1) is stable. If

$$\dot{V}(t) \leq -\omega_3(|x(t)|) \quad (7.5)$$

then the solution $x(t) = 0$ is asymptotically stable.

Proof. Assume $\varepsilon > 0$. Take $\delta(\varepsilon)$ such that $\omega_1(\varepsilon) = \omega_2(\delta)$. From (7.3), (7.4) it follows for $\|\varphi(\theta)\|_W \leq \delta(\varepsilon)$ that

$$\omega_1(\|x_t\|_W) \leq V(t) \leq V(t_0) \leq \omega_2(\|\varphi\|_W) \leq \omega_2(\delta) = \omega_1(\varepsilon)$$

This implies that $|x(t)|^2 \leq \|x_t\|_W^2 \leq \varepsilon^2$; i.e., $|x(t)| \leq \varepsilon$. This proves stability. To prove asymptotic stability, choose H_1 such that $\omega_1(H) = \omega_2(H_1)$. Then, as previously demonstrated, $\|\varphi\|_W \leq H_1$ and inequality (7.2) implies $\|x_t\|_W \leq H$ and $|\dot{x}(t)| \leq M$. Suppose that a solution $x(t, t_0, \varphi)$, $\|\varphi\|_W \leq H_1$ does not tend to zero as $t \to \infty$. Then for some $\varepsilon > 0$ there exists a sequence $t_i \to \infty$, $i \to \infty$, such that $|x(t_i, t_0, \varphi)| \geq \varepsilon$. Therefore, on the intervals $t_i - \Delta \leq t \leq t_i + \Delta$, $\Delta = \varepsilon(2M)^{-1}$ we have $|x(t, t_0, \varphi)| \geq \varepsilon/2$ and $\dot{V}(t) \leq -\omega_3(\varepsilon/2) = \gamma < 0$. Therefore,

$$V(t) - V(t_0) = \int_0^t \dot{V}(s)\, ds \geq \sum_{t_i \leq t} \int_{t_i - \Delta}^{t_i + \Delta} \dot{V}(s)\, ds \geq 2\gamma\, \Delta N(t)$$

where $N(t)$ is a number of points t_i such that $t_i < t$. Thus $V(t) - V(t_0) > -\infty$ as $t \to \infty$. This contradicts inequality (7.3). The proof of Theorem 7.1 is complete.

7.2. Examples

Consider the Eq. [150]

$$\dot{x}(t) = -ax(t) + c(t)\dot{x}(t - h), \qquad t \geq 0 \tag{7.6}$$

where $c(t)$ is a continuous function. Define the functional [150]

$$V(t, x_t, \dot{x}_t) = x^2(t) + \frac{1}{a} \int_{t-h}^t \dot{x}^2(s)\, ds$$

For $\dot{V}(t)$ we get

$$\dot{V}(t) = 2x(t)\dot{x}(t) + a^{-1}[\dot{x}^2(t) - \dot{x}^2(t - h)]$$

Replacing $\dot{x}(t)$ according to (7.6) we obtain for $|c(t)| < 1$ that

$$\dot{V}(t) = -ax^2(t) - a^{-1}[1 - c^2(t)]\dot{x}^2(t - h) \leq -ax^2(t)$$

Therefore, the functional $V(t, x_t, \dot{x}_t)$ satisfies all the conditions of Theorem 7.1. This implies the asymptotic stability of the solution $x(t) = 0$. For other examples see El'sgol'tz and Norkin [59] and Misnik [150].

Castelan and Infante [35], for a matrix autonomous NFDE

$$\dot{x}(t) = Ax(t) + Bx(t - h) + C\dot{x}(t - h), \qquad x(t) \in R_n \tag{7.7}$$

have constructed the functional $V(t, x_t, \dot{x}_t)$ satiisfying the conditions of Theorem 7.1 under the assumption that the trivial solution of (7.7) is asymptotically stable. A Liapunov functional was obtained as the limit, in an appropriate sense, of a Liapunov function constructed by well-known methods for a difference equation approximation of the original NFDE (7.7).

§8. STABILITY AND BOUNDEDNESS OF LINEAR NONHOMOGENEOUS EQUATIONS

8.1. Connection between Stability and Boundedness

Consider the NFDE

$$\dot{x}(t) = L(t, x_t, \dot{x}_t) + f(t), \quad t > 0$$
$$x_0(\theta) = \varphi(\theta), \quad \dot{x}(\theta) = \dot{\varphi}(\theta), \quad -h \leq \theta \leq 0 \quad (8.1)$$

Here the operator $L(t, x_t, \dot{x}_t) = A_1(t, \dot{x}_t) + A_2(t, x_t)$ is linear in the second and third arguments. Assume that for problem (8.1) conditions of existence, uniqueness and continuous dependence of solutions on initial data are met.

Theorem 8.1 [253(5)]. *If any solution of problem* (8.1) *with* $f(t) = 0$ *is bounded for* $t \geq 0$, *then its trivial solution is stable.*

Notice that, in general, the boundedness of all the solutions of a nonlinear equation does not imply their stability (see Example 2.1.1).

Let the operators $A_1(t, \dot{x}_t)$ and $A_2(t, x_t)$ be bounded and ω periodic in t. Consider functional equation

$$z(t) = A_1(t, z_t) + g(t), \quad t > 0, \quad z(\theta) = 0, \quad \theta \leq 0 \quad (8.2)$$

Problem (8.2) is said to *satisfy the Perron condition* if its solution defined by

Theorem 8.2 [162(13)]. *Let operator* $L(t, x_t, \dot{x}_t)$ *be linear in* x_t, $\dot{x}_t$ *and periodic in* t. *Assume that problem* (8.2) *satisfies the Perron condition and that Eq.* (8.1) *has a bounded solution for the functions* $f(t) \in L_\infty[0, \infty)$ *and* $\varphi(\theta) = 0$. *Then the trivial solution of Eq.* (8.1) *is asymptotically stable.*

8.2. Boundedness of Derivatives

Consider an ordinary differential equation with bounded coefficients

$$x^{(m)}(t) + A_1(t)x^{(m-1)}(t) + \cdots + A_m(t) \times (t) = f(t)$$
$$x^{(m)} = d^m x/dt^m \quad (8.3)$$

It is known the following *Esclangon's theorem* [36].

Let there exist a bounded solution $x(t)$, $-\infty < t < \infty$ of Eq. (8.3) for some bounded function $f(t)$. Then all derivatives $x(t), \ldots, x^{(m)}(t)$, $-\infty < t < \infty$, are bounded.

This theorem, extended on differential-difference equations in [194], is not valid in general for NFDEs. Consider the scalar NFDE ($2\pi \leq t$)

$$\dot{x}(t) - \dot{x}(\sqrt{t^2 - 2}) = 2(t - \sqrt{t^2 - 2\pi})x(\sqrt{t^2 - 3\pi/2})$$

The coefficient of this equation is bounded for $t \geq 2\pi$ and $f(t) = 0$. But this equation has bounded solution $x(t) = \sin(t^2)$, $t \geq 2\pi$, with unbounded derivative.

Prove an Esclangon-type theorem for the NFDE

$$\begin{aligned} x^{(m)}(t) + A_1(t, x_t^{(m)}) + A_2(t, x_t, \ldots, x_t^{(m-1)}) = f(t), \quad t > 0 \\ x(\theta) = \dot{x}(\theta) = \cdots = x^{(m)}(\theta) = 0, \quad -h \leq \theta \leq 0 \end{aligned} \quad (8.4)$$

Here continuous operators $A_1: R_1 \times L_\infty[-h, 0] \to R_n$, $A_2: R_1 \times C[-h, 0] \times \cdots \times C[-h, 0] \to R_n$ are linear in all arguments beginning from the second. In addition, $A_1(t, \varphi) \in L_\infty[0, T]$ for any $\varphi(s) \in L_\infty[-h, T]$, $T > 0$, and $A_2(t, \psi_t^{(1)}, \ldots, \psi_t^{(m)}) \in L_\infty[0, T]$ for any $\psi^{(m)}(s) \in C[-h, T]$. Assume also that

$$\operatorname*{vrai\,sup}_{0 \leq t \leq T} |A_2(t, \psi_t^{(1)}, \ldots, \psi_t^{(m)})|$$

$$\leq a[\|\psi^{(1)}\|_{C[-h, T]} + \cdots + \|\psi^{(m)}\|_{C[-h, T]}) \quad (8.5)$$

Condition (8.5) replaces the assumption about the boundedness of the Eq. (8.3) coefficients. Consider the functional equation

$$z(t) + A_1(t, z_t) = R(t), \quad t > 0; \quad z(\theta) = 0, \quad -h \leq \theta \leq 0$$

$$R(t) \in L_\infty[0, \infty] \quad (8.6)$$

Theorem 8.3 [108(5)]. *Let operators A_1 and A_2 satisfy the formulated requirements and Perron condition be met for Eq. (8.6). If there exists a bounded solution $x(t)$ of Eq. (8.4) for a bounded function $f(t) \in L_\infty[0, \infty)$, then $\operatorname{vrai\,sup}_{0 \leq t < \infty}[|\dot{x}(t)| + \cdots + |x^{(m)}(t)|] < \infty$.*

§9. LINEAR PERIODIC EQUATIONS

9.1. Some Examples

Linear periodic NFDEs are studied less than RFDEs because the monodromy operator for NFDEs is not completely continuous. We now illustrate by examples some features of periodic NFDEs.

§9. Linear Periodic Equations

EXAMPLE 9.1. The Floquet multiplier of NFDE may have *infinite multiplicity*. For the equation $\dot{x}(t) - \dot{x}(t-1) = (\cos 2\pi t)[x(t) - x(t-1)]$, the Floquet multiplier $\rho = 1$ has infinite multiplicity since any continuously differentiable 1-periodic function $x(t)$ is its solution.

EXAMPLE 9.2. Consider a scalar equation which *has no Floquet solution*

$$\dot{x}(t) - \dot{x}(t - 2\pi) = (\cos t)x(t - 2\pi) \tag{9.1}$$

If $x(t) = \exp(\lambda t)y(t)$ is a Floquet solution of (9.1), then $y(t)$ must be a 2π-periodic solution of the equation

$$\dot{y}(t)[1 - \exp(-2\pi\lambda)] = [(\cos t)\exp(-2\pi\lambda) - \lambda(1 - \exp(-2\pi\gamma))]y(t) \tag{9.2}$$

But for any complex number λ Eq. (9.2) has no nontrivial periodic solutions. Let be $\lambda = mi$, $m = 0, \pm 1, \pm 2, \ldots$. Then the unique solution of Eq. (9.2) having the form $0 = (\cos t)y(t)$ is $y(t) = 0$. If $\lambda \neq mi$, $m = 0, \pm 1, \ldots$, then the general solution of Eq. (9.2) is

$$y(t) = C \exp\left\{ \int_0^t \left[\frac{\exp(-2\pi\lambda)}{1 - \exp(-2\pi\lambda)} \cos s - \lambda \right] ds \right\}$$

This solution will be 2π periodic if and only if

$$\int_0^{2\pi} \left[\frac{\exp(-2\pi\lambda)}{1 - \exp(-2\pi\gamma)} \cos s - \lambda \right] ds = -2\pi\lambda = 0$$

But it is impossible in the domain $\lambda \neq mi$, $m = 0, \pm 1, \ldots$. Hence, there are no nontrivial periodic solutions of Eq. (9.2) and the Floquet solutions of Eq. (9.1).

EXAMPLE 9.3. Consider an equation whose Floquet solutions *do not give an asymptotic representation for other solutions*:

$$2\dot{x}(t) - \dot{x}(t - 2\pi) = (\cos t)x(t - 2\pi) \tag{9.3}$$

Research of Floquet solutions of Eq. (9.3) is equivalent to the determination of a λ such that the equation

$$\dot{y}(t)[2 - \exp(-2\pi\lambda)] = [(\cos t)\exp(-2\pi\lambda) - \lambda(2 - \exp(-2\pi\lambda))]y(t) \tag{9.4}$$

has a nontrivial period solution. If $[2 - \exp(-2\pi\lambda)] = 0$, then Eq. (9.4) has no nontrivial periodic solutions. If $[2 - \exp(-2\pi\lambda)] \neq 0$, then the periodic solution of Eq. (9.4) exists only for $\lambda = 0$. Then this periodic solution $y(t)$ is simultaneously a Floquet solution $x_0(t)$ of Eq. (9.3) $x_0(t) = y(t) = \exp(\sin t)$. Now find solution $x_1(t)$ of Eq. (9.3) with initial data $x_1(t) = \sin t$, $\dot{x}_1(t) = \cos t$, $0 \le t \le 2\pi$. By the step method we have

$$x_1(t) = \sum_{k=1}^{m+1} A_m^k (\sin t)^k, \quad 2m\pi \le t \le 2\pi(m+1), \quad m = 0, 1, \ldots \tag{9.5}$$

Here A_m^k are defined by the formulae

$$A_m^k = A_{m-1}^k/2 + A_{m-1}^{k-1}/2, \qquad A_0^1 = 1, \quad A_m^0 = A_m^{m+2} = 0 \qquad (9.6)$$

From (9.6) it follows that $A_m^k > 0$, $m = 1, 2, \ldots$; $1 \le k \le m+1$ and

$$\max_{2\pi m \le t \le 2\pi(m+1)} |x_1(t)| = x_1(t_m) = \sum_{k=1}^{m+1} A_m^k, \qquad t_m = \frac{\pi}{2} + 2\pi m$$

$$\frac{1}{2} \sum_{k=1}^{m} A_{m-1}^k \le \sum_{k=1}^{m+1} A_m^k \le \frac{3}{4} \sum_{k=1}^{m} A_{m-1}^k \qquad (9.7)$$

So

$$\frac{1}{2} x_1(t_{m-1}) \le x_1(t_m) \le \frac{3}{4} x_1(t_{m-1}) \qquad (9.8)$$

By virtue of (9.7) and (9.8) we get $|x_1(t)| \le (\frac{3}{4})^{(t/2\pi)-1}$. Hence $\lim x_1(t) = 0$, $t \to \infty$. But $\inf_t x_0(t) > 0$. Consequently, the function $Cx_0(t)$ is the best approximation of solution $x_1(t)$ if and only if $C = 0$. The obtained Floquet representation is not asymptotic for the solution $x_1(t)$. From (9.8) it follows that $x_1(t_m) \ge 2^{-m}$. Therefore the difference between the solution $x_1(t)$ and its Floquet representation $Cx_0(t) = 0$ decreases less quickly than does $\exp[-\ln 2(t/2\pi - \frac{1}{4})]$. So the Floquet solution $x_0(t)$ does not give an asymptotic representation for $x_1(t)$ since β from Theorem 2.12.2 must be equal or greater than $(-\ln 2/2\pi)$.

Examples 9.1–9.3 show that a full generalization of Floquet theory for NFDEs is impossible. In particular, Theorem 2.12.2 is not valid for NFDEs. Notice, however, that a system of Floquet solutions is complete in the space $C[-q\omega, 0]$, $q = \max(m, n)$ for the scalar equation

$$\dot{x}(t) = a\dot{x}(t - m\omega) + q(t)x(t - n\omega), \qquad |a| < 1$$

Here the function $q(t) \ge 0$ does not equal zero in whole intervals [116].

9.2. Stability Theorem

Consider the equation

$$(d/dt)[x(t) - G(t, x_t)] = L(t, x_t) \qquad (9.9)$$

Here operators $G: R_1 \times C[-h, 0] \to R_n$ and $L: R_1 \times C[-h, 0] \to R_n$ are continuous in both arguments, linear in the second argument and ω periodic in t. The operator $G(t, \varphi)$ is called a *stable* one if the trivial solution of the homogeneous functional equation $x(t) - G(t, x_t) = 0$, $t \ge 0$, is uniformly asymptotically stable.

§9. Linear Periodic Equations

Theorem 9.1 [81(4), 108(5)]. *Let the previously formulated assumption be met and let the operator G be stable. Then there exists a number $a \in (0, 1)$ such that Eq. (9.9) has a finite number of Floquet multipliers ρ_j for which $|\rho_j| \geq a$. The trivial solution of Eq. (9.9) is stable if and only if all multipliers ρ_j satisfy the condition $|\rho_j| \leq 1$ and those with $|\rho_k| = 1$ have simple elementary Jordan divisors. Further, the trivial solution of Eq. (9.9) is asymptotically stable if and only if $|\rho_j| < 1$.*

Chapter 4

Stability of Stochastic Functional Differential Equations

§1. SOME PREREQUISITES FROM THE THEORY OF STOCHASTIC RETARDED FUNCTIONAL DIFFERENTIAL EQUATIONS

1.1. Statement of the Initial-Value Problem

This chapter is devoted to the study of the theory of SRFDEs

$$dx(t) = a(t, x_t) \, dt + b(t, x_t) \, d\xi(t), \qquad t \geq 0$$
$$x_t = x(t + \theta), \qquad -\infty < \theta \leq 0, \qquad x(t) \in R_n \qquad (1.1)$$

Here $\xi(t) \in R_1$ is a *standard Wiener process*, defined on the probability space $(\Omega, \mathfrak{I}, P)$. Recall that the elements ω are called elementary events, and the sets in $\mathfrak{I}$ are called random events (see [64, 84, 108(9)]). The probability measure P defined on the σ algebra $\mathfrak{I}$ is a nonnegative countably additive set function on $\mathfrak{I}$ such that $P(\Omega) = 1$. A random variable $\zeta(\omega) \in R_n$ is a function defined on Ω such that $\{\omega : \zeta(\omega) < x\} \in \mathfrak{I}$ for any $x \in R_n$. The family of random variables $\zeta(t, \omega) \in R_n$, $t \geq 0$ is called a stochastic process. For any fixed $\omega \in \Omega$ the function $\zeta(t, \omega)$ considered as a function of t is called a trajectory or sample function of the stochastic process. The stochastic process ζ determines the minimal σ algebra whose related random variables $\zeta(s, \omega)$ are measurable for $s \leq t$. Usually the argument ω of the stochastic process $\zeta(t, \omega)$ is omitted. The standard Wiener process $\xi(t)$ has independent stationary gaussian increments and

$$\xi(0) = 0, \qquad E[\xi(t) - \xi(s)] = 0$$
$$E[\xi(t) - \xi(s)][\xi(t) - \xi(s)]' = I(t - s)$$

where E denotes expectation, I is a unit matrix and "prime" is a transportation sign.

Let $B_{t_1 t_2}(d\xi) \subset \mathfrak{I}$ be a minimal σ algebra whose related random variables $\xi(s) - \xi(t)$ are measurable for any $t_1 \leq t \leq s \leq t_2$. The sample functions of

§1. Some Prerequisites from the Theory of SRFDEs

the process $\xi(t)$ are continuous, nowhere differentiable, have infinite variations on any finite time interval and satisfy Holder's condition with an index less than $\frac{1}{2}$. The upper limit of Wiener process samples equals $+\infty$ with probability 1 for $t \to \infty$ and the lower limit equals $-\infty$.

The continuous functionals $a(t, \varphi)$ and $b(t, \varphi)$ in Eq. (1.1) are determined on $[0, \infty) \times CB[-\infty, 0]$. Given the stochastic process $\varphi(\theta)$, $\theta \le 0$, the initial condition for (1.1) is

$$x(\theta) = \varphi(\theta), \qquad \theta \le 0 \qquad (1.2)$$

Assume that σ algebra $B_{-\infty 0}(\varphi)$ is independent from $B_{0\infty}(d\xi)$. The stochastic process $x(t)$ is called a *solution of initial-value problem* (1.1), (1.2) if $B_{t\infty}(d\xi)$ is independent from $B_{-\infty t}(x) \cup B_{0t}(d\xi)$ for any $t \ge 0$ and with probability 1 we will have

$$x(t) = x(0) + \int_0^t a(s, x_s)\, ds + \int_0^t b(s, x_s)\, d\xi(s) \qquad (1.3)$$

where $x(\theta) = \varphi(\theta)$ for $\theta \le 0$. The last integral in (1.3) is Itô's stochastic integral having the following properties [64, 84]:

$$\int_0^T [c_1 f_1(t) + c_2 f_2(t)]\, d\xi(t) = c_1 \int_0^T f_1(t)\, d\xi(t) + c_2 \int_0^T f_2(t)\, d\xi(t)$$

$$E \int_0^T f_1(t)\, d\xi(t) = 0,$$

$$E \int_0^T f_1(t)\, d\xi(t) \int_0^T f_2(s)\, d\xi(s) = \int_0^T E f_1(t) f_2(t)\, dt$$

Here the arbitrary processes $f_1(t)$, $f_2(t)$, $t \in [0, T]$ are measurable relative to σ algebra $B_{-\infty 0}(\varphi) \cup B_{0t}(d\xi)$ and

$$\int_0^T E|f_1(t)|^2\, dt < \infty$$

1.2. Existence Theorem

Assume that for any $x(\theta)$, $y(\theta) \in CB[-\infty, 0]$ and $t \ge 0$

$$|a(t, x) - a(t, y)|^2 \le \int_0^\infty |x(-s) - y(-s)|^2\, dR_1(s)$$

$$|b(t, x) - b(t, y)|^2 \le \int_0^\infty |x(-s) - y(-s)|^2\, dR_2(s) \qquad (1.4)$$

$$|b(t, x)|^2 = \sum_{i=1}^n \sum_{j=1}^l b_{ij}^2(t, x)$$

where R_1 and R_2 are nondecreasing functions with bounded variations

$$r_i^2 = \int_0^\infty dR_i(s) < \infty, \qquad i = 1, 2$$

It is always assumed that the process $\varphi(\theta)$ has continuous samples and

$$\sup_{\theta \leq 0} E|\varphi(\theta)|^4 < \infty \tag{1.5}$$

Theorem 1.1 [91]. *Let Eq. (1.1) and the initial function satisfy the conditions formulated above. Then there exists a unique solution of problem (1.1), (1.2) with bounded fourth moment on any finite time interval. This solution is measurable relative to the processes $\varphi(\theta)$, $\theta \leq 0$ and $\xi(s)$, $0 \leq s \leq t$, i.e., $B_{-\infty t}(x) \subset B_{-\infty 0}(\varphi) \cup B_{0t}(d\xi)$. The process x_t is Markovian.*

1.3. Itô's Formula

Itô's formula will be useful in the investigation of the stability of system (1.1). Given a scalar continuous function $u(t, x)$, $t \geq 0$, $x \in R_n$ with continuous partial derivatives $u_t(t, x)$, $u_x(t, x) = (\partial u / \partial x_i)$, $u_{xx}(t, x) = (\partial^2 u / \partial x_i \partial x_j)$, $i, j = 1, \ldots, n$. Then the stochastic differential $d\eta(t)$ of the process $\eta(t) = u[t, x(t)]$ [where $x(t)$ is a solution of Eq. (1.1)] equals

$$\begin{aligned} d\eta(t) = \{&u_t[t, x(t)] + u_x'[t, x(t)]a(t, x_t) \\ &+ \tfrac{1}{2} \operatorname{Tr} b(t, x_t)b'(t, x_t)u_{xx}[t, x(t)]\} \, dt \\ &+ u_x'[t, x(t)]b(t, x_t) \, d\xi(t) \end{aligned} \tag{1.6}$$

Here $\operatorname{Tr} bb'$ is a trace of the matrix bb'.

1.4. Equations for Moments

Consider the linear SRFDE

$$dx(t) = \int_0^\infty [d_s A(t, s)] x(t - s) \, dt + \int_0^\infty [d_s B(t, s)] x(t - s) \, d\xi(\varepsilon) \tag{1.7}$$

with initial condition (1.2). Here A and B are $(n \times n)$ matrix functions. From Itô's formula we obtain the deterministic equations for the moments of solution to SRFDE (1.7). For $m(t) = Mx(t)$ we get

$$\dot m(t) = \int_0^\infty [d_s A(t, s)] m(t - s), \quad m(0) = E\varphi(0)$$

§2. Formulation of Stability Problems for SRFDEs

and for $k(t, s) = Ex(t)x'(s)$, we get

$$\frac{dk(t,t)}{dt} = \int_0^\infty \{k(t, t-s)\, d_s A'(t, s)$$
$$+ [d_s A(t, s)]k(t-s, t)\}$$
$$+ \int_0^\infty \int_0^\infty [d_s B(t, s)]k(t-s, t-s_1)[d_{s_1} B'(t, s_1)]$$

Let Eq. (1.7) be of the form

$$dx(t) = a(t)x(t-h)\, dt + b(t)\, d\xi(t), \qquad t \geq 0, \quad h \geq 0$$

Then $J = Ex'(T)Hx(T)$ (where $T > 0$ and a matrix H are given) is equal to

$$J = E\bigg[\varphi'(0)P\varphi(0) + \varphi'(0)\int_{-h}^0 Q(s)\varphi(s)\, ds$$
$$+ \int_{-h}^0 \varphi'(s)Q'(s)\varphi(0)\, ds$$
$$+ \int_{-h}^0 \int_{-h}^0 \varphi'(s)R(s, \rho)\varphi(\rho)\, ds\, d\rho\bigg] + \alpha$$

Here

$$P = B_1'(0)HB_1(0), \qquad Q = B_1'(0)H\dot{B}_1(s)$$
$$R(s, \rho) = \dot{B}_1'(s)H\dot{B}_1'(\rho)$$
$$\alpha = \int_0^T \text{Tr}\, B_1'(t)HB_1(t)b(t)b'(t)\, dt$$

The matrix B_1 is easily calculated by the step method from the equation

$$\dot{B}_1(t) = -B_1(t+h)a(t+h), \qquad B_1(T) = I, \quad B_1(s) = 0, \quad s > T$$

§2. FORMULATION OF STABILITY PROBLEMS FOR SRFDEs. LIAPUNOV DIRECT METHOD

2.1. Basic Definitions of Stability

In this paragraph various definitions and theorems relating to stochastic stability are formulated for vector SRFDEs (1.1). It is always assumed that Eq. (1.1) satisfy requirements from §1 and also that

$$a(t, 0) = 0, \qquad b(t, 0) = 0 \qquad (2.1)$$

From condition (2.1) it follows that Eq. (1.1) has a trivial solution $x(t) = 0$ defined by a zero initial condition. The stability of the trivial solution of system (1.1) is investigated relative to the disturbances of the initial function $\varphi(\theta)$. Denote $x(t, \varphi)$ the solution of problem (1.1), (1.2).

Definition 2.1. *The trivial solution of SRFDE* (1.1) *is called:* (a) *p stable if for each $\varepsilon > 0$ there exists $\delta(\varepsilon) > 0$ such that for any initial process $\varphi(\theta)$ with continuous trajectories and independent from $B_{0\infty}(d\xi)$, the inequalities*

$$\sup_{\theta \leq 0} E|\varphi(\theta)|^p < \delta(\varepsilon), \qquad \sup_{\theta \leq 0} E|\varphi(\theta)|^{2p} < \infty \qquad (2.2)$$

imply that $E|x(t, \varphi)|^p < \varepsilon$, $t \geq 0$; (b) *asymptotically p stable if it is p stable and for any arbitrary initial function $\varphi(\theta) \in Q$*

$$\lim_{t \to \infty} E|x(t, \varphi)|^p = 0 \qquad (2.3)$$

The domain Q is called an attraction domain of the trivial solution.

If $p = 2$ then the definition 2.1 is a definition of *mean-square stability*.

Definition 2.2. *The trivial solution of SRFDE* (1.1) *is called:* (a) *stable with respect to probability if for each $\varepsilon_1 > 0$ and $\varepsilon_2 > 0$ there exists a $\delta > 0$ such that the solution $x(t, \varphi)$ of problem (1.1), (1.2) satisfies the inequality*

$$P\left\{\sup_{t \geq 0} |x(t, \varphi)| \leq \varepsilon_1\right\} \geq 1 - \varepsilon_2$$

provided that $\sup_{\theta \leq 0} |\varphi(\theta)| \leq \delta$ with probability 1; (b) *asymptotically stable with respect to probability if it is stable with respect to probability and for any positive number λ_1 and λ_2 and arbitrary initial function $\varphi(\theta) \in S_H$ with probability 1, there exists $T(H, \lambda_1, \lambda_2) > 0$ such that*

$$P\left\{\sup_{t \geq T(H, \lambda_1, \lambda_2)} |x(t, \varphi)| \leq \lambda_1\right\} \geq 1 - \lambda_2$$

Definition 2.3. *The trivial solution of SRFDE* (1.1) *is called exponentially p stable ($p > 0$) if for any positive constants c_1 and c_2*

$$E|x(t, \varphi)|^p \leq c_1 \sup_{\theta \leq 0} E|\varphi(0)|^p \exp(-c_2 t), \qquad t \geq 0$$

2.2. Asymptotic *p* Stability

General theorems of the Liapunov direct method for SRFDEs may be obtained by formal generalization of the appropriate theorems for

§2. Formulation of Stability Problems for SRFDEs

deterministic equations from Chapter 2, §2. Let us formulate, therefore, the only theorem used in this section.

Theorem 2.1. *Let Eq. (1.1) satisfy requirements from §1 and (2.1). Assume that on $[0, \infty) \times CB[-\infty, 0]$ there exists the continuous functional $V(t, \varphi)$ such that*

$$V(t, \varphi) \geq c_1 |\varphi(0)|^p, \qquad t \geq 0, p \geq 2 \tag{2.4}$$

$$EV(t, \varphi) \leq c_2 \sup_{\theta \leq 0} E|\varphi(\theta)|^p, \tag{2.5}$$

$$EV[t_2, x_{t_2}(\varphi)] - EV[t_1, x_{t_1}(\varphi)] \leq -c_3 \int_{t_1}^{t_2} E|x(t)|^p \, dt \tag{2.6}$$

Then the trivial solution of Eq. (1.1) is asymptotically p stable.

Proof. From (2.6) it follows that

$$EV(t, x_t(\varphi)) \leq EV(0, \varphi) - c_3 \int_0^t E|x(s, \varphi)|^p \, ds$$

Taking into account (2.4), (2.5) we get

$$c_1 E|x(t, \varphi)|^p \leq c_2 \sup_{\theta \leq 0} E|\varphi(\theta)|^p \tag{2.7}$$

Hence, p stability is proved. Justify (2.3). By (2.7) and Itô's formula we conclude that $E|x(t, \varphi)|^p$ as a function of t satisfies the Lipschitz condition and

$$\int_0^\infty E|x(t, \varphi)|^p \, dt < \infty$$

Therefore relation (2.3) holds true.

REMARK 2.1. Theorem 2.1 is valid if, instead of inequality (2.6), the following inequality holds:

$$dV(t, x_t) \leq -c_3 |x(t)|^p \, dt + a_1(t) \, d\xi(t), \qquad t > 0$$

Here the continuous process $a_1(t)$ is measurable relative to $B_{-\infty 0}(\varphi) \cup B_{0t}(d\xi)$ and has a bounded second moment on any finite time interval.

2.3. Exponential p Stability

Formulate the necessary and sufficient conditions [230(2)] of exponential p stability for a stationary SRFDE with finite delay

$$dx(t) = a(x_t)\, dt + b(x_t)\, d\xi(t), \qquad t \geq 0$$
$$x(t + 0) = x_t, \qquad x_0(\theta) = \varphi(\theta), \qquad -h \leq \theta \leq 0 \tag{2.8}$$

Theorem 2.2. *The necessary and sufficient condition of exponential p stability of the trivial solution of Eq. (2.8) is the existence of functional $V(\varphi)$ with the properties*

$$c_1 \|\varphi\|^p \leq V(\varphi) \leq c_2 \|\varphi\|^p$$

$$\varlimsup_{t \to +0} \frac{1}{t} \{EV[x_t(\varphi)] - V(\varphi)\} \leq -c_3 \|\varphi\|^p$$

REMARK 2.2. Theorems 2.1 and 2.2 are valid when the stochastic disturbance $\xi(t)$ is a general process with independent increments

2.4. SRFDEs with Random Delays [99, 129]

Consider the system

$$\dot{x}(t) = Ax(t) + Bx[t - \eta(t)], \qquad t \geq 0$$
$$x_0(\theta) = \varphi(\theta), \qquad -h \leq \theta \leq 0, \qquad x \in R_n, \qquad \varphi(\theta) \in C[-h, 0] \tag{2.9}$$

Here A and B are constant matrices and $\eta(t) \in [0, h]$ is a homogeneous scalar Markovian process with a finite number of states $\eta(0) = \eta_0$. Here η_0 is any point from the interval $[0, h]$.

Theorem 2.3. *Let the continuous functional $V(\varphi, \eta)$ exist such that, uniformly in η,*

$$\omega_1(\|\varphi\|) \leq V(\varphi, \eta) \leq \omega_2(\|\varphi\|), \qquad \eta \in [0, h]$$

$$\varlimsup_{t \to +0} \frac{1}{t} [EV[x_t(\varphi), \eta(t)] - V(\varphi, \eta_0)] \leq -\omega_3(\|\varphi\|), \qquad \eta(0) = \eta_0$$

Then the trivial solution of Eq. (2.9) is stable with respect to probability.

Theorem 2.3 permits us to relate the stability of SRFDE (2.9) to the stability of the corresponding deterministic equations. Let, for example, the deterministic system (2.9) with $\eta(t) = \eta$ be exponentially stable uniformly in η for arbitrary $\eta \in [0, h]$ and also the derivative

$$dE|\eta(t) - \eta_0|/dt \tag{2.10}$$

be sufficiently small. Then the trivial solution of (2.9) is stable with respect to probability.

Notice that in general this statement is not valid if the derivative (2.10) is not small (see Example 2.1.4).

§3. STABILITY OF SCALAR STOCHASTIC EQUATIONS

3.1. Preliminaries

Derive some sufficient conditions for mean-square stability of solutions of the scalar SRFDE

$$dx(t) = \left[-\int_0^\infty x(t-s) \, dK_0(s) + a(t, x_t) \right] dt$$
$$+ b(t, x_t) \, d\xi(t), \qquad t > 0 \qquad (3.1)$$
$$x(\theta) = \varphi(\theta), \qquad -\infty < \theta \le 0, \qquad x(t) \in R_1$$

Here the kernel $K_0(s)$ has a bounded variation on $[0, \infty)$. Continuous functionals $a(t, \varphi)$ and $b(t, \varphi)$ are determined on the space $[0, \infty) \times CB[-\infty, 0]$ and satisfy conditions (1.4), (2.1). The initial process $\varphi(\theta)$ satisfies the assumptions from §1.

3.2. Case of Autonomous Linear Part

Full proofs of the theorems in this section are available [105(5), 105(11), 108(5)]. Here we limit our presentation only by the form of the Liapunov functionals and stochastic differentials used in the proofs. Denote

$$\beta_{10} = \int_0^\infty s^i \, dK_0(s), \qquad \alpha_{i0} = \int_0^\infty s^i |dK_0(s)|$$

$$\int_0^\infty s \, dR_1(s) + \int_0^\infty s \, dR_2(s) < \infty$$

Theorem 3.1. *Let Eq. (3.1) satisfy the requirements from §1 and kernel $K_0(s)$ have a jump at zero of value $a_0 > 0$ and*

$$\beta_1 = a_0 - \int_{+0}^\infty dK_0(s) - r_1 - \frac{1}{2} r_2^2 > 0, \qquad \alpha_{10} < \infty$$

Then the trivial solution of Eq. (3.1) is asymptotically mean-square stable.

Introduce the functional

$$V(t, x_t) = x^2(t) + \int_0^\infty \left[dK_0(s) + r_1^{-1} dR_1(s) \right.$$
$$\left. + dR_2(s) \right] \int_{t-s}^t x^2(t_1) dt_1 \qquad (3.2)$$

The stochastic differential of functional (3.2), calculated with the aid of Itô's formula, is equal to

$$dV(t, x_t) = \left\{ b^2(t, x_t) + x^2(t) \left[\int_{+0}^\infty dK_0(s) | + r_1 + r_2^2 \right] \right.$$
$$- \int_{+0}^\infty x^2(t-s) |dK_0(s)| + 2x(t) \left[- \int_0^\infty x(t-s) dK_0(s) \right.$$
$$\left. + a(t, x_t) \right] \right\} dt + 2x(t) b(t, x_t) d\xi(t)$$
$$+ \int_0^\infty x^2(t-s) [r_1^{-1} dR_1(s) + dR_2(s)]$$

By virtue of (1.4) and (2.1)

$$-2x(t) \int_0^\infty x(t-s) dK_0(s)$$
$$= -2a_0 x(t) - 2x(t) \int_{+0}^\infty x(t-s) dK_0(s)$$
$$\leq -2a_0 x(t) + x^2(t) \int_{+0}^\infty |dK_0(s)| + \int_{+0}^\infty x^2(t-s) |dK_0(s)|$$
$$2x(t) a(t, x_t) \leq r_1 x^2(t) + r_1^{-1} a^2(t, x_t)$$
$$\leq r_1 x^2(t) + r_1^{-1} \int_0^\infty x^2(t-s) dR_1(s)$$

Hence
$$dV(t, x_t) \leq -2x^2(t) \beta_1 \, dt + 2x(t) b(t, x_t) \, d\xi(t) \qquad (3.3)$$

The validity of Theorem 3.1 now follows from (3.2), (3.3), Theorem 2.1 and the obvious estimations

$$V(t, x_t) \geq x^2(t)$$

$$EV(t, x_t) \leq \sup_{\theta \leq 0} Ex^2(t+\theta) \left[1 + \alpha_{10} + \int_0^\infty s \, dR_2(s) + r_1^{-1} \int_0^\infty s \, dR_1(s) \right]$$

§3. Stability of Scalar Stochastic Equations

For the scalar equation

$$dx(t) = -ax(t)\,dt + bx(t-h)\,d\xi(t), \qquad t \geq 0, \quad h \geq 0$$

Theorem 3.1 gives that the trivial solution is asymptotically mean-square stable if $2a > b^2$.

Theorem 3.2. *Let conditions (1.4), (1.5) are (2.1) be met, let the kernel $K_0(s)$ be non-decreasing and let*

$$\alpha_{10} < 1, \qquad \beta_2 = (1-\alpha_{10})\beta_{00} - (1+\alpha_{10})r_1 - \tfrac{1}{2}r_2^2 > 0, \qquad \alpha_{20} < \infty$$

Then the trivial solution of Eq. (3.1) is asymptotically mean-square stable.

Consider the functional

$$V(t, x_t) = V_0(t, x_t) + \int_0^\infty dR_2(s) \int_{t-s}^t x^2(t_1)\,dt_1 \tag{3.4}$$

where

$$V_0 = \left[x(t) - \int_0^\infty dK_0(s) \int_{t-s}^t x(t_1)\,dt_1 \right]^2 + (\alpha_{00} + r_1) \int_0^\infty dK_0(s)$$

$$\cdot \int_{t-s}^t dt_1 \int_{t_1}^t x^2(t_2)\,dt_2 + (1+\alpha_{10})r_1^{-1} \int_0^\infty dR_1(s) \int_{t-s}^t x^2(t_1)\,dt_1$$

For the stochastic differential of functional (3.4) the following relation is valid:

$$dV(t, x_t) \leq -\beta_2 x^2(t)\,dt + 2\left[x(t) - \int_0^\infty dK_0(s) \int_{t-s}^t x(t_1)\,dt_1 \right] b(t, x_t)\,d\xi(t) \tag{3.5}$$

Theorem 3.2 follows from (3.4), (3.5) and Theorem 2.1.

Theorem 3.3. *Let conditions (1.4), (1.5), (2.1) be fulfilled and let*

$$\alpha_{10} < 1, \qquad \alpha_{20} < \infty, \qquad (1-\alpha_{10})\beta_{00} > (1+\alpha_{10})r_1 + \tfrac{1}{2}r_2^2$$

Then the trivial solution of Eq. (3.1) is asymptotically mean-square stable.

The proof of this theorem is based on consideration of the functional

$$V(t, x_t) = \left[x(t) - \int_0^\infty dK_0(s) \int_{t-s}^t x(t_1)\, dt_1 \right]^2$$
$$+ (\beta_{00} + r_1) \int_0^\infty dK_0(s) \int_{t-s}^t dt_1 \int_{t_1}^t x^2\, dt_2$$
$$+ \int_0^\infty [(1 + \alpha_{10})r_1^{-1}\, dR_1(s) + dR_2(s)\mu] \int_{t-s}^t x^2(t_1)\, dt_1 \quad (3.6)$$

For

$$dx(t) = [ax(t) - bx(t - h)]\, dt + cx(t)\, d\xi(t), \qquad t \geq 0 \quad (3.7)$$

from Theorem 3.2 it follows that the trivial solution of Eq. (3.7) is asymptotically mean-square stable if

$$0 < bh < 1, \qquad 2b(1 - bh) > c^2 + 2(1 + bh)|a|$$

Note that the trivial solution of Eq. (3.7) with $b = 0$, $a > 0$ is mean-square unstable. Consequently it is possible to stabilize a system by introducing time lag.

3.3. Case of Nonautonomous Linear Part

Some modifications of functionals (3.2), (3.4), (3.6) permit us to investigate Eq. (3.1) where kernel $K_0(s)$ is exchanged on $K_0(t, s)$. Consider, for example, the equation

$$dx(t) = \left[a(t, x_t) + \int_0^\infty x(t - s) f(t, s)\, ds - \alpha_1(t) x(t) \right] dt$$
$$+ b(t, x_t)\, d\xi(t), \qquad t \geq 0 \quad (3.8)$$

The functionals $a(t, x_t)$ and $b(t, x_t)$ in (3.8) are the same as in Eq. (3.1), the function $\alpha_1(t)$ is continuous and bounded for $t \in [0, \infty)$, the function $f(t, s)$ is continuous for all t, s and

$$\int_{s_1}^{s_2} |f(t, s)|\, ds \leq q(s_2) - q(s_1), \qquad s_2 \geq s_1$$

where $q(s)$ is a bounded nondecreasing function. The mean-square asymptotic stability of Eq. (3.8) takes place under the conditions

$$\inf_{t \geq 0}\left[2\alpha_1(t) - r_2^2 - 2r_1 - \int_0^\infty |f(t + s, s)|\, ds - \int_0^\infty |f(t, s)|\, ds \right] > 0$$
$$\sup_{t \geq 0}\left[\int_0^\infty ds \int_t^{t+s} |f(t_1, s)|\, dt_1 \right] < \infty \quad (3.9)$$

§4. Stability of Second-Order Equations

These stability conditions are proved with the aid of the functional

$$V(t, x_t) = x^2(t) + \int_0^\infty ds \int_{t-s}^t |f(t_1 + s, s)| x^2(t_1) \, dt_1$$
$$+ \int_0^\infty [r_1^{-1} \, dR_1(s) + dR_2(s)] \int_{t-s}^t x^2(t_1) \, dt_1$$

For the equation

$$dx(t) = -a(t)x(t) + bx(t - h) \, d\xi(t)$$
$$t > 0, \qquad h \geq 0, \qquad b \text{ a const}$$

where the function $a(t)$ is continuous and bounded, the stability condition is $2a(t) > b^2$, $t \geq 0$.

§4. STABILITY OF SECOND-ORDER EQUATIONS

4.1. General Assumptions

Obtain some conditions for the mean-square stability of the second-order SRFDE

$$\dot{x}(t) = y(t), \qquad t > 0$$
$$dy(t) = \left[-\int_0^\infty y(t-s) \, dK_0(s) - \int_0^\infty x(t-s) \, dK_1(s) \right] dt$$
$$+ b(t, y_t) \, d\xi(t), \qquad y_t = y(t + \theta), \quad -\infty < \theta \leq 0 \quad (4.1)$$

The coefficients of Eq. (4.1) will always be assumed to satisfy the conditions from §1. The kernels K_1, K_0 have bounded variations on $[0, \infty)$ and the integrals in (4.1) are Riemann–Stieltjes ones [105(11)]. Let, for the arbitrary functions $\varphi_1, \varphi_2 \in CB[-\infty, 0]$,

$$|b(t, \varphi_1) - b(t, \varphi_2)|^2 \leq \int_0^\infty |\varphi_1(-s) - \varphi_2(-s)|^2 \, dK_2(s) \quad (4.2)$$

and let

$$b(t, 0) = 0 \quad (4.3)$$

where $K_2(s)$ is a nondecreasing function of bounded variation on $[0, \infty)$. Under these assumptions there exists a unique solution $[x(t), y(t)]$ of system (4.1) satisfying the initial conditions

$$x(\theta) = \varphi(\theta), \qquad y(\theta) = \dot{\varphi}(\theta), \qquad \theta \leq 0 \quad (4.4)$$

Here $\varphi(\theta)$ is a given process with continuously differentiable realization such that

$$\sup_{\theta \leq 0} E[\varphi^4(\theta) + \dot{\varphi}^4(\theta)] \tag{4.5}$$

It is assumed that σ algebra $B_{-\infty 0}(\varphi, \dot{\varphi})$ is independent from the σ algebra $B_{0\infty}(d\xi)$. In this case the inclusion $B_{-\infty t}(x, y) \subset [B_{-\infty 0}(\varphi, \dot{\varphi}) \cup B_{0t}(d\xi)]$ is valid. From (4.1)–(4.5) it follows that for all $t \geq 0$ the solution of problem (4.1) satisfies the inequality

$$E[x^4(t) + y^4(t)] \leq c_1 \sup_{\theta \leq 0} E[\varphi^4(\theta) + \dot{\varphi}^4(\theta)] e^{c_2 t}, \qquad c_i \geq 0 \tag{4.6}$$

4.2. Stability Conditions

The conditions formulated in this section are expressed in terms of moments of the kernels $K_j(s)$. Denote

$$\alpha_{ij} = \int_0^\infty s^i |dK_j(s)|, \qquad \beta_{ij} = \int_0^\infty s^i \, dK_j(s)$$

$$r(t) = E[x^2(t) + y^2(t)], \qquad \gamma(\varphi) = \sup_{\theta \leq 0} E[\varphi^2(\theta) + \dot{\varphi}^2(\theta)]$$

Theorem 4.1. *Let the coefficients of system (4.1) satisfy requirements (4.1), (4.2), and let the kernel $K_0(s)$ have a jump at zero of value $a > 0$, with*

$$\beta_{01} > 0, \, a > \int_{+0}^\infty |dK_0(s)| + \alpha_{11} + \alpha_{02}, \qquad \alpha_{10} + \alpha_{21} + \alpha_{12} < \infty$$

Then the trivial solution of system (4.1) is asymptotically mean-square stable.

Proof. Introduce the functional

$$V(x_t, y_t) = 2\beta_{01} x^2(t) + y^2(t) + \int_{+0}^\infty |dK_0(s)| \int_{t-s}^t [y^2(t_1)$$

$$+ \beta_{01} x^2(t_1)] \, dt_1 + 2 \int_0^\infty dK_2(s) \int_{t-s}^t y^2(t_1) \, dt_1$$

$$+ \int_0^\infty |dK_1(s)| \int_{t-s}^t dt_1 \int_{t_1}^t [y^2(t_2) + \beta_{01} x^2(t_2)] \, dt_2$$

$$+ V_0^2(x_t, y_t) \tag{4.7}$$

where

$$V_0(x_t, y_t) = y(t) + \int_0^\infty x(t-s) \, dK_0(s) - \int_0^\infty dK_1(s) \int_{t-s}^t x(t_1) \, dt_1 \tag{4.8}$$

§4. Stability of Second-Order Equations

On the basis of (4.1) and Itô's formula we have

$$dy^2(t) = b^2(t, y_t) \, dt + 2y(t)b(t, y_t) \, d\xi(t)$$
$$+ 2y(t)\left[-ay(t) - \int_{+0}^{\infty} y(t-s) \, dK_0(s)\right.$$
$$\left. - \int_0^{\infty} x(t-s) \, dK_1(s)\right] dt \qquad (4.9)$$

But by (4.2), (4.3) with probability 1

$$b^2(t, y_t) \leq \int_0^{\infty} y^2(t-s) \, dK_2(s) \qquad (4.10)$$

Moreover

$$2y(t) \int_{+0}^{\infty} y(t-s) \, dK_0(s) \leq y^2(t) \int_{+0}^{\infty} |dK_0(s)|$$
$$+ \int_{+0}^{\infty} y^2(t-s)|dK_0(s)| \qquad (4.11)$$

$$-2y(t) \int_0^{\infty} x(t-s) \, dK_1(s)$$
$$= -2y(t)x(t)\beta_{01} + 2y(t) \int_0^{\infty} dK_1(s)_{t-s} y(t_1) \, dt_1$$
$$\leq -2y(t)x(t)\beta_{01} + \alpha_{11}y^2(t) + \int_0^{\infty} |dK_1(s)| \int_{t-s}^t y^2(t_1) \, dt_1 \quad (4.12)$$

From (4.9)–(4.12) we get

$$dy^2(t) \leq \left[\int_0^{\infty} x^2(t-s) \, dK_2(s) + \int_{+0}^{\infty} y^2(t-s)|dK_0(s)|\right.$$
$$-2x(t)y(t)\beta_{01} + y^2(t)\left(-2a + \alpha_{11} + \int_{+0}^{\infty} |dK_0(s)|\right) \qquad (4.13)$$
$$\left. + \int_0^{\infty} |dK_1(s)| \int_{t-s}^t y^2(t_1) \, dt_1\right] dt + 2y(t)b(t, y_t) \, d\xi(t)$$

It should be noted that inequality (4.13) means that for arbitrary $t_2 \geq t_1 \geq 0$ the integral in the left side of (4.13) from t_1 to t_2 is, with probability 1, less than or equal to the integral in the right side of (4.13) between the same limits. From the second equation of Eqs. (4.1) and Itô's formula we obtain

$$dV_0^2(x_t, y_t) = 2V_0(x_t, y_t)[-x(t)\beta_{01} \, dt + b(t, y_t) \, d\xi(t)]$$
$$+ b^2(t, y_t) \, dt \qquad (4.14)$$

Estimating as in (4.11) the terms on the right side of (4.14) we obtain according to (4.8)

$$dV_0^2(x_t, y_t) \leq 2V_0(x_t, y_t)b(t, y_t)\,d\xi(t) + \beta_{01}\Big\{-2x(t)y(t)$$

$$+ x^2(t)\bigg[\int_{+0}^{\infty} |dK_0(s)| - 2a + \alpha_{11}\bigg] + \int_{+0}^{\infty} x^2(t-s)|dK_0(s)|$$

$$+ \int_0^{\infty} |dK_1(s)| \int_{t-s}^{t} x^2(t_1)\,dt_1\bigg\}\,dt$$

$$+ \int_0^{\infty} x^2(t-s)\,dK_2(s)\,dt \tag{4.15}$$

From (4.8), (4.13) and (4.15) follows the formula

$$dV(x_t, y_t) \leq -2x^2(t)\beta_{01}\bigg[a - \int_{+0}^{\infty} |dK_0(s)| - \alpha_{11}\bigg]$$

$$+ 2b(t, y_t)[V_0(x_t, y_t) + y(t)]\,d\xi(t)$$

$$- 2y^2(t)\bigg[a - \int_{+0}^{\infty} |dK_0(s)| - \alpha_{11} - \alpha_{02}\bigg] \tag{4.16}$$

Let us integrate inequality (4.16) from zero to $t \geq 0$ and take the expectation on both sides of the result. By (4.5), (4.6), (4.10) and the hypotheses of Theorem 4.1, the expectation of the stochastic integral in the right side of (4.16) is equal to zero. Hence for all $t \geq 0$

$$EV(x_t, y_t) - EV(\varphi, \dot{\varphi}) \leq -\int_0^t [\gamma_1 Ey^2(s) + \gamma_2 Ex^2(s)]\,ds \tag{4.17}$$

Here

$$\gamma_1 = 2\bigg[a - \int_{+0}^{\infty} |dK_0(s)| - \alpha_{11} - \alpha_{02}\bigg]$$

$$\gamma_2 = 2\beta_{01}\bigg[a - \int_{+0}^{\infty} |dK_0(s)| - \alpha_{11}\bigg] \tag{4.18}$$

But by the hypotheses of Theorem 4.1 together with (4.8)

$$EV(\varphi, \dot{\varphi}) \leq c_1\gamma(\varphi) \tag{4.19}$$

$$EV(x_t, y_t) \leq \beta_{01}Ex^2(t) + Ey^2(t) \tag{4.20}$$

From these relations and (4.17) it follows that $r(t) \leq c_2\gamma(\varphi)$ for arbitrary $t \geq 0$. So the mean-square stability is established. For the proof of asymptotic

§4. Stability of Second-Order Equations   179

mean-square stability it suffices to show that the function $r(t)$ is uniformly continuous on $[0, \infty)$ and

$$\int_0^\infty r(t)\, dt < \infty \tag{4.21}$$

But inequality (4.21) follows immediately from (4.17)–(4.20), since they imply that for all $t \geq 0$

$$\int_0^t [\gamma_1 Ey^2(s) + \gamma_2 Ex^2(s)]\, ds \leq c_1\gamma(\varphi)$$

Let us prove now that the function $r(t)$ is uniformly continuous on $[0, \infty)$. For this we apply Itô's formula to the function $x^2(t) + y^2(t)$. From (4.10), (4.11) we obtain that for arbitrary nonnegative t_1 and t_2

$$|r(t_1) - r(t_2)| \leq 2c_2\gamma(\varphi)(1 + \alpha_{00} + \alpha_{01} + \alpha_{02})|t_1 - t_2|$$

Theorem 4.1 is proved

For the equation ($t \geq 0$)

$$\dot{x}(t) = y(t), \qquad dy(t) = [-a_1 y(t) + a_2 x(t - h_1)]\, dt + a_3 y(t - h_2)\, d\xi(t)$$

Theorem 4.1 gives that the trivial solution is asymptotically mean-square stable if $a_2 > 0$ and $a_1 > a_2 h_1 + a_3^2$.

Theorem 4.2. *Let the conditions* (4.2), (4.3) *be satisfied and let*

$$q_1 = \alpha_{10} + \tfrac{1}{2}\alpha_{21} < 1, \qquad \beta_{01} > 0$$

$$q = \beta_{00} - \beta_{11} > \max[q_1, (\beta_{01} q_1 + \alpha_{02})/(1 - q_1)]$$

$$\alpha_{11} + \alpha_{21} + \alpha_{20} + \alpha_{31} < \infty$$

Then the trivial solution of system (4.1) *is asymptotically mean-square stable.*

The proof of this theorem is based on the functional

$$V(x_t, y_t) = V_0^2(x_t, y_t) + V_1^2(y_t) + 2x^2(t)\beta_{01}$$

$$+ 2\int_0^\infty dK_2(s) \int_{t-s}^t y^2(t_1)\, dt_1$$

$$+ (q + 2\beta_{01})\left[\int_0^\infty |dK_0(s)| \int_{t-s}^t dt_1 \int_{t_1}^t y^2(t_2)\, dt_2\right.$$

$$\left. + \int_0^\infty |dK_1(s)| \int_{t-s}^t dt_1 \int_{t_1}^t dt_2 \int_{t_2}^t y^2(t_3)\, dt_3\right]$$

where V_0 is given by (4.8) and

$$V_1(y_t) = y(t) - \int_0^\infty dK_0(s) \int_{t-s}^t y(t_1) \, dt_1$$
$$+ \int_0^\infty dK_1(s) \int_{t-s}^t dt_1 \int_{t_1}^t y(t_2) \, dt_2$$

By Itô's formula we obtain

$$dV(x_t, y_t) \leq -2x^2(t)\beta_{01}(q - q_1) \, dt$$
$$+ 2b(t, y_t)[V_1(y_t) + V_0(x_t, y_t)] \, d\xi(t)$$
$$- 2y^2(t)[-\beta_{01}q_1 + q(1 - q_1) - \alpha_{02}] \, dt \quad (4.23)$$

The validity of Theorem 4.2 follows from relations (4.22), (4.23) (for details see Kolmanovskii [105(7), 105(11), and Kolmanovskii and Nosov 108(5)]].)

REMARK 4.1. The results of Chapter 2, §6 and Chapter 3, §4 show that it is not difficult to alter functionals (4.7), (4.22) in such a way that one can obtain the conditions of asymptotic mean-square stability of the nonautonomous Eq. (4.1).

REMARK 4.2. Applying to functionals (4.7), (4.22) a slight modification of the arguments in Has'minskii [84] one finds, under the hypotheses of Theorems 4.1 and 4.2, not only mean-square stability, but also stability with respect to probability in the sense of Definition 2.2.

4.3. Stability Domains of Linear Autonomous Systems Obtained by the Liapunov Direct Method

Some sufficient stability conditions for second order SRFDEs are derived by use of quadratic Liapunov functionals by Shaichet [199(1), 199(2)]. Consider the equation

$$dx(t) = [ay(t) - bx(t - h)] \, dt + c_1 x(t) \, d\xi_1(t)$$
$$dy(t) = [-ax(t) - by(t - h)] \, dt + c_2 y(t) \, d\xi_2(t), \quad t > 0 \quad (4.24)$$

Sufficient asymptotic mean-square stability conditions for Eq. (4.24) are

$$|a| < [h^{-2}(1 - sb^{-1})^2 - b^2]^{1/2}, \quad s = c^2/2, \quad c = max(|c_1|, |c_2|).$$

Stability domains of (4.24) for different values of parameter sh are represented in Fig. 4.1. The case in which $h = 0$ is represented in Fig. 4.2.

For the equations

$$dx(t) = [ay(t) - bx(t - h)] \, dt + c_1 x(t) \, d\xi_1(t),$$
$$dy(t) = [ax(t) - bx(t - h)] \, dt + c_2 y(t) \, d\xi_2(t), \quad t > 0 \quad (4.25)$$

§4. Stability of Second-Order Equations

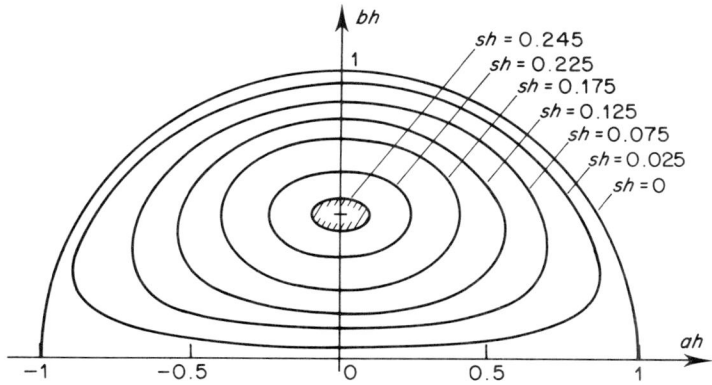

Fig. 4.1. Stability domains for system (4.24).

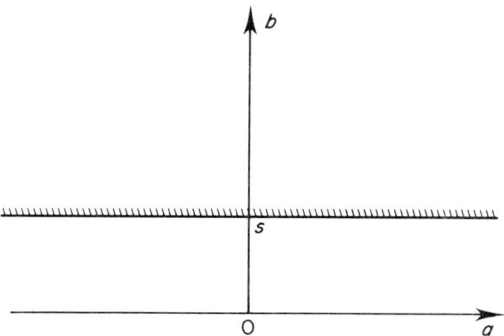

Fig. 4.2. Degenerated case of $h = 0$.

asymptotic mean-square stability takes place if $|a| < [b(1 - bh) - s]$ $\cdot (1 + bh)^{-1}$. Corresponding stability domains are shown in Figs. 4.3 and 4.4. Note that for $a = 0$ the boundary points of stability domains for Eqs. (4.24) and (4.25) are $b_{1,2} = (2h)^{-1}[1 \pm (1 - 4sh)^{1/2}]$. If $sh \geq 0.25$ then the stability domains degenerate.

In Figs. 4.1 and 4.3 stability domains are situated inside corresponding curves.

Consider the equation

$$\ddot{x}(t) + [a^2 + c\dot{\xi}(t)]x(t) + 2b\dot{x}(t - h) + b^2 x(t - 2h) = 0 \qquad (4.26)$$

Sufficient asymptotic mean-square stability conditions of Eq. (4.26) are

$$c^2/2 = s < a^2 b[1 - h(a^2 + b^2)^{1/2}], \qquad b > 0 \qquad (4.27)$$

182   4. Stability of Stochastic Functional Differential Equations

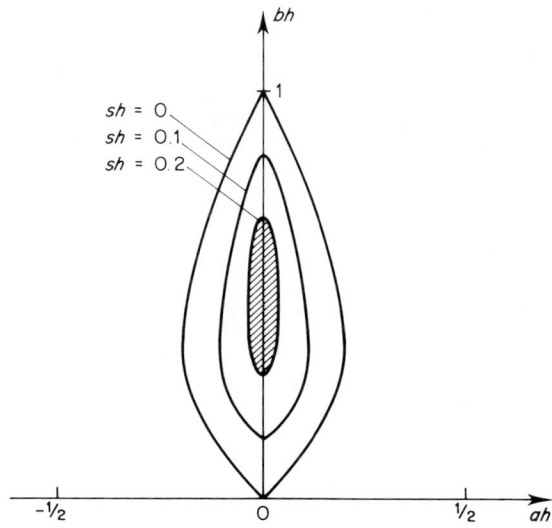

Fig. 4.3. Stability domains for system (4.25).

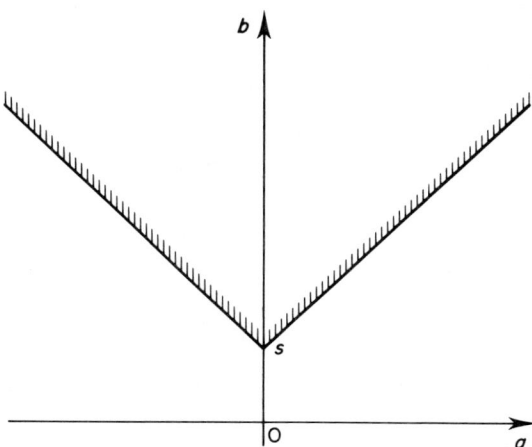

Fig. 4.4. Degenerated case of $h = 0$.

§4. Stability of Second-Order Equations

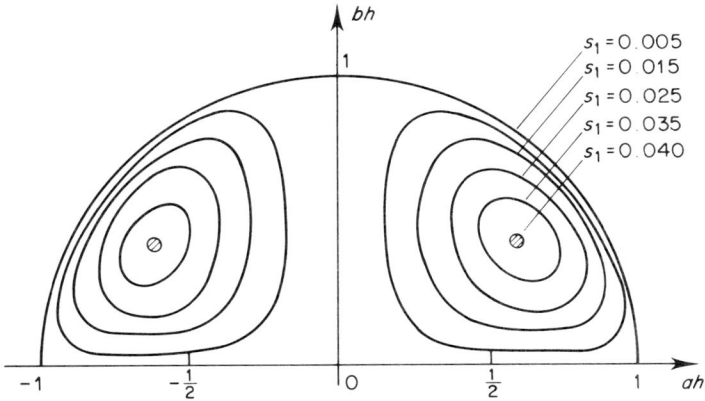

Fig. 4.5. Stability domains for system (4.26).

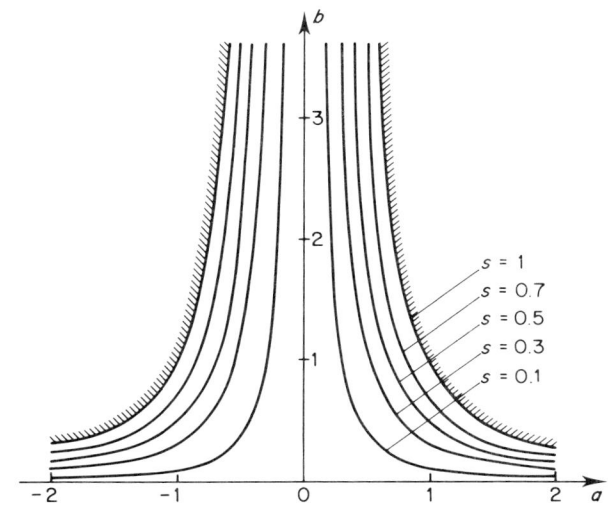

Fig. 4.6. Degenerated case of $h = 0$, $s > 0$.

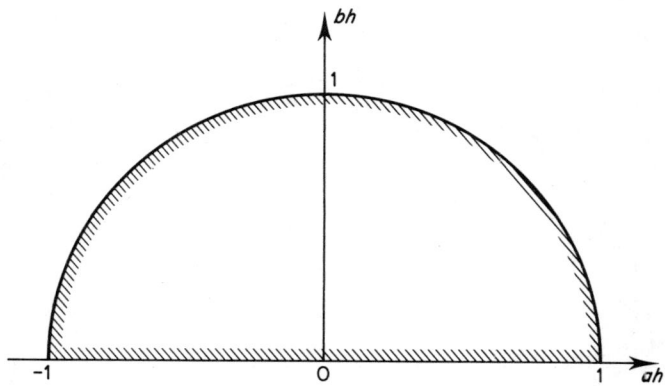

Fig. 4.7. Degenerated case of $s = 0$, $h > 0$.

Inequality (4.27) is impossible if $s_1 = sh^3 > 3\sqrt{3}/128 \cong 0.040595$. The stability domain is shown in Fig. 4.5. Degenerated cases in which $h = 0$, $s = 0$ or $s = h = 0$ are shown in Fig. 4.6–4.8.

For the equations

$$\ddot{x}(t) - [a^2 + c\dot{\xi}(t)]x(t) + 2b\dot{x}(t - h) + b^2 x(t - 2h) = 0 \qquad (4.28)$$

the stability domains represented in Fig. 4.9 are given by the inequality $c^2/2 = s < a^2 b[1 - h(|a| + |b|)] - |a|^3$, $b > 0$. This inequality is impossible for $sh^3 > (219\sqrt{73} - 1871)/128 \cong 0.001069$. Degenerated cases are shown in Fig. 4.10–4.12.

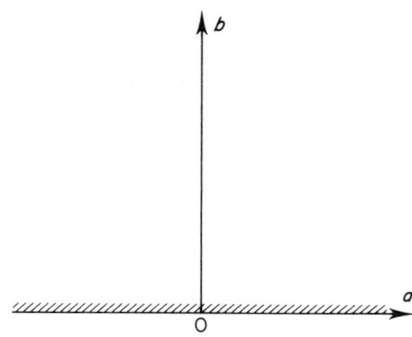

Fig. 4.8. Case of $h = s = 0$.

§4. Stability of Second-Order Equations   185

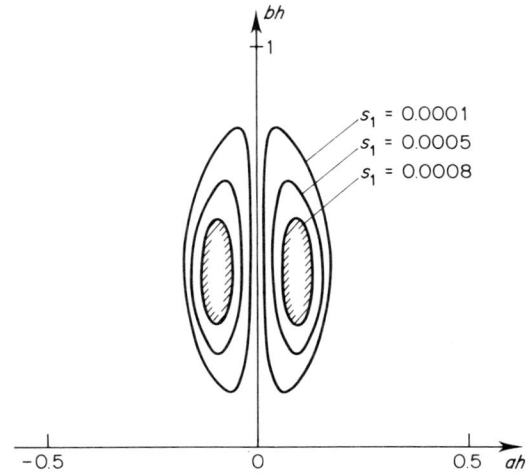

Fig. 4.9. Stability domains for Eq. (4.28).

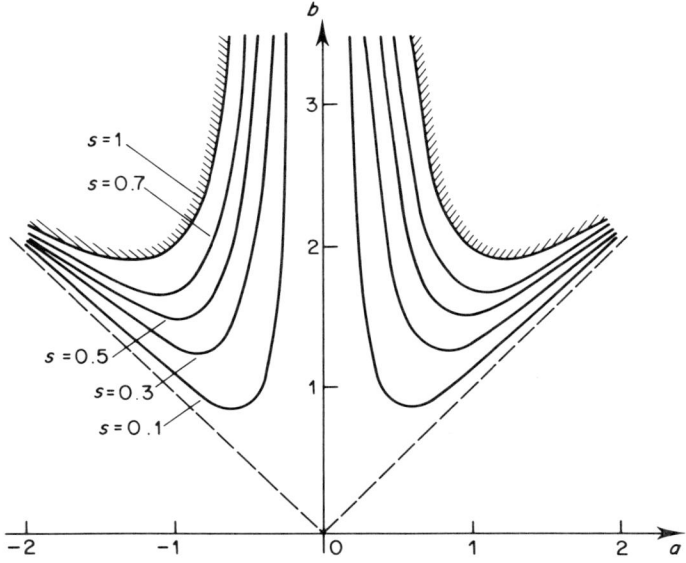

Fig. 4.10. Degenerated case ($h = 0, s > 0$).

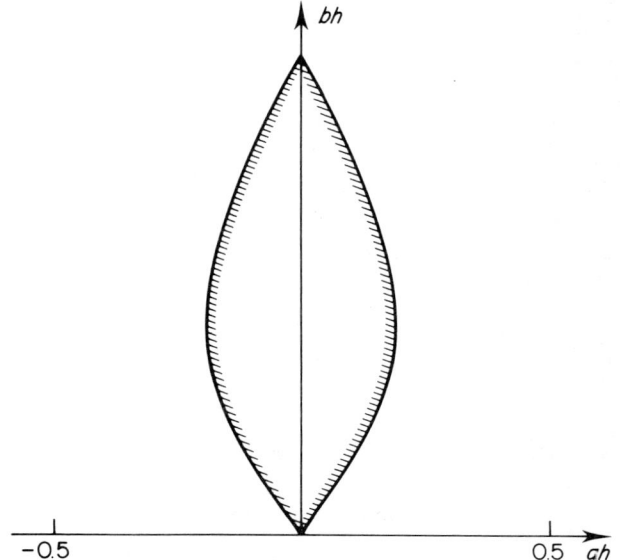

Fig. 4.11. Degenerated case ($s = 0$, $h > 0$).

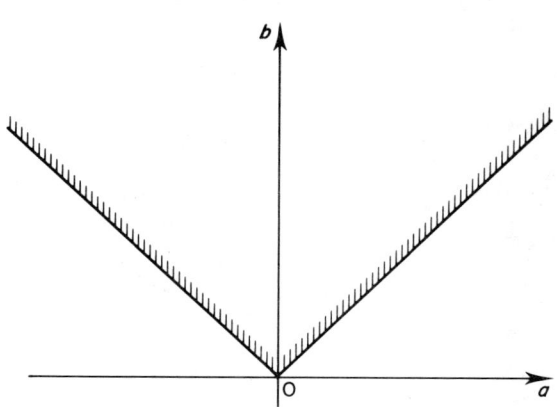

Fig. 4.12. Case of $s = h = 0$.

§5. STATIONARY AND PERIODIC SOLUTIONS OF STOCHASTIC RETARDED EQUATIONS

This section is devoted to the stationary and periodic solutions of SRFDEs. Some results about existence of stationary solutions are derived on the basis of the Liapunov direct method.

5.1. Existence of Stationary Solutions

Let $\zeta(t) \in R_n$ be a continuous stationary (in the strict sense) process, i.e., for every sequence $(t_1, \ldots, t_N)$ the joint probability distribution of the variables $\zeta(t_1 + T), \ldots, \zeta(t_N + T)$ is independent of T. Consider initial-value problem

$$\dot{x}(t) = a[x_t, \zeta(t)], \quad t > 0, \quad x \in R_n \tag{5.1}$$

$$x(\theta) = \varphi(\theta), \quad -h \leq \theta \leq 0, \quad \varphi(\theta) \in C[-h, 0] \tag{5.2}$$

Here continuous functional $a(\varphi, z) \in R_n$ is such that

$$|a(\varphi_1, z) - a(\varphi_2, z)| \leq r\|\varphi_1 - \varphi_2\|, \quad \varphi_1, \varphi_2 \in C[-h, 0]$$

$$P\left\{\int_0^t |a[0, \zeta(s)]|\,ds < \infty\right\} = 1, \quad \forall t > 0$$

Definition 5.1. *A stationary process $x(t)$ is a stationary solution of Eq. (5.1) if $x(t)$ is stationary connected with $\zeta(t)$ and satisfies Eq. (5.1).*

Theorem 5.1. *Let there exist such a solution $\bar{x}(t)$ of (5.1), (5.2) such that*

(1) *uniformly in $t > 0$*

$$\lim_{c \to \infty} \frac{1}{t} \int_0^t P\{\bar{x}(s) > C\}\, ds = 0$$

(2) *for any positive $\varepsilon_1, \varepsilon_2, C$ and $t_1, t_2 \in [-h, \infty)$*

$$E|x(t_1) - x(t_2)|^{\varepsilon_1} \leq C|t_1 - t_2|^{1+\varepsilon_2} \tag{5.3}$$

Then there exists a stationary solution of Eq. (5.1).

For the proof of Theorem 5.1 see Kolmanovskii [105(1) and Kolmanovskii and Nosov 108(5)].

5.2. Relation between Stability and Stationarity of Solutions

Let $a(\varphi, \zeta) = a_1(\varphi) + \zeta(t)$, $a_1(0) = 0$ in Eq. (5.1), i.e.,

$$\dot{x}(t) = a_1(x_t) + \zeta(t), \quad t > 0 \tag{5.4}$$

The existence of a stationary solution of Eq. (5.4) depends on the stability "degree" of the deterministic system

$$\dot{y}(t) = a_1(y_t), \qquad t \geq 0 \tag{5.5}$$

Theorem 5.2. *Assume that the trivial solution of Eq. (5.5) is uniformly exponentially stable and also that $E|\zeta(t)|^2 < \infty$. Then there exists a stationary solution of Eq. (5.4).*

Proof. Consider the solution $x(t)$ of (5.4) with zero initial condition (5.2). Show that $E|x(t)|^2 \leq c < \infty$. From the conditions of Theorem 5.2 follow the existence of the continuous functional $V(\varphi)$ (Chapter 2, §5) such that

$$c_1 \|\varphi\| \leq V(\varphi) \leq c_2 \|\varphi\|, \qquad \varphi \in C[-h, 0]$$

$$|V(\varphi) - V(\varphi_1)| \leq c_3 \|\varphi - \varphi_1\|, \qquad \varphi_1 \varphi_1 \in C[-h, 0] \tag{5.6}$$

$$\dot{V}(x_t) \leq -c_4 \|x_t\|$$

Let the function $z(t)$ be determined by

$$\dot{z}(t) = a_1(z_t), \qquad t \geq s; \qquad z(t) = x(t), \qquad t \leq s$$

Then for any $s \geq 0$

$$\dot{V}(x_s) = \overline{\lim_{\Delta \to +0}} \, (1/\Delta)[V(z_{s+\Delta}) - V(x_s)]$$

$$+ \overline{\lim_{\Delta \to +0}} \, (1/\Delta)[V(x_{s+\Delta}) - V(z_{s+\Delta})] \tag{5.7}$$

$$\leq -(c_4/c_2)V(x_s) + c_3|\zeta(s)|$$

Hence, applying Lemma 1.4.2 we get

$$V(x_s) \leq c_3 \int_0^s \exp[-c_5(s-t)]|\zeta(t)|\,dt$$

From (5.7) and (5.6) it follows that

$$c_1 \|x_s\| \leq c_3 \int_0^s \exp[-c_5(s-t)]|\zeta(t)|\,dt$$

§5. Stationary and Periodic Solutions

So

$$E\|x_s\|^2 \leq E[c_6 \int_0^s \exp[-c_5(s-t)]|\zeta(t)|\,dt]^2$$

$$\leq c_6^2 \left\{ \int_0^\infty \exp[-c_5(t-s)]\,dt \right\}$$

$$\cdot \left\{ \int_0^\infty \exp[-c_5(t-s)] E|\zeta(t)|^2\,dt \right\} \leq c_7 \quad (5.8)$$

From Chebyshev's inequality [64] we obtain

$$\frac{1}{t}\int_0^t P\{|x(s)| > c\}\,ds \leq c_7 c^{-2} \to 0 \quad \text{for} \quad c \to \infty \quad (5.9)$$

By virtue of (5.4), (5.8)

$$E|x(t+\Delta) - x(t)|^2$$

$$\leq 2\Delta \int_t^{t+\Delta} E[a_1^2(x_s) + \zeta^2(s)]\,ds$$

$$\leq 2\Delta \int_t^{t+\Delta} E[r^2\|x_s\|^2 + \zeta^2(s)]\,ds \leq c_8 \Delta^2 \quad (5.10)$$

Theorem 5.2 now follows from Theorem 5.1 and relations (5.9), (5.10).

5.3. Periodic Solutions

A stochastic process $\beta(t) \in R_m$ is said to be *periodic with period* T if for every sequence $(t_1, \ldots, t_N)$ the joint probability distribution of the stochastic variables $\beta(t_1 + mT), \ldots, \beta(t_N + mT)$ (where $m = \pm 1 \pm 2, \ldots$) is independent of m. Consider the equation

$$\dot{x}(t) = b[t, x(t), \beta(t)], \quad t \geq 0, \quad x \in R_n, \quad \beta \in R_m \quad (5.11)$$

Here the continuous functional $b(t, \varphi, z)$ for any fixed φ and z is periodic in t with period T and also

$$|b(t, \varphi, z) - b(t, \psi, z)| \leq c\|\varphi - \psi\|, \quad \forall\, \varphi, \psi \in C[-h, 0]$$

$$P\left\{ \int_0^t |b[s, 0, \eta(s)]|\,ds < \infty \right\} = 1, \quad \forall\, t \geq 0, \quad \forall\, z \in R_n \quad (5.12)$$

Definition 5.2. *A periodic process $x(t)$ is called a periodic solution of Eq. (5.11) if $x(t)$ is periodically connected with $\beta(t)$ and satisfies to Eq. (5.11).*

Theorem 5.3. *Let conditions* (5.3) *and* (5.12) *hold and there exist at least one solution* $x(t)$ *of Eq.* (5.11) *such that uniformly in* k

$$\lim_{c \to \infty} \frac{1}{k} \sum_{i=0}^{k-1} P\{|\bar{x}(iT)| > c\} = 0, \qquad k = 1, 2, \ldots$$

Then there exists a periodic solution of Eq. (5.11).

The proof can be found in Kolmanovskii [105(1)] and Kolmanovskii and Nosov [108(5)].

5.4. Ergodic Properties of Stationary Solutions

We present here some ergodic properties of stationary solutions of the SRFDE.

$$dx(t) = a_1(x_t) \, dt + b[x(t)] \, d\xi(t), \qquad t \geq 0$$
$$x(t) \in R_n, \qquad x_t = x(t + \theta), \qquad -h \leq \theta \leq 0 \qquad (5.13)$$

Ergodic properties of Eq. (5.13) solutions are investigated by means of stationary solutions of a stochastic equation without delay

$$dy(t) = a_2[y(t)] \, dt + b[y(t)] \, d\xi(t), \qquad t \geq 0 \qquad (5.14)$$

Here $\xi(t)$ is a standard Wiener process, the functional $a_1(\varphi)$, $\varphi \in C[-h, 0]$ and functions $a_2(x)$, $b(x)$ are continuous and satisfy the Lipschitz conditions. Any stationary solution $x_t(\varphi)$ of Eq. (5.13) with initial condition (5.2) induces the invariant measure μ on σ algebra $\mathfrak{I}_0$ of Borel sets from $C[-h, 0]$.

Definition 5.3. *Stationary solution* $x_t(\varphi)$ *of Eq.* (5.13) *is called ergodic if for an arbitrary* $\mathfrak{I}_0$ *measurable functional* $\lambda(\varphi)$, $\varphi \in C[-h, 0]$,

$$\lim_{t \to \infty} \frac{1}{t} \int_0^t \lambda[x_s(\varphi)] \, ds = \int_{C[-h, 0]} \lambda(\varphi) \mu(d\varphi)$$

almost everywhere with respect to invariant measure μ.

Theorem 5.4. *Let there exist the unique stationary ergodic solution of Eq.* (5.14). *If there exists stationary solution of Eq.* (5.13) *then it is unique and ergodic.*

The proof of Theorem 5.4, which is available in Kolmanovskii [105(4)], is based on absolute continuity of measures induced by solutions of Eqs. (5.13) and (5.14). In Kolmanovskii [105(4)], and Kolmanovskii and Nosov [108(5)] the wide-sense mixing property of stationary solutions of SDFE (5.13) are also obtained.

§6. STABILITY WITH RESPECT TO THE FIRST APPROXIMATION

6.1. General Theorem

Conditions of stability theorems with respect to the first approximation depend on the form of the first approximation equations. Establish a theorem for the case in which the equation of the first approximation is a deterministic one

$$\dot{y}(t) = a(t, y_t), \qquad t > 0, \quad y \in R_n$$
$$y_t = y(t + \theta), \qquad -h \leq \theta \leq 0 \tag{6.1}$$

with initial conditions (1.2). Continuous functionals $a(t, \varphi)$ and $b(t, \varphi)$ satisfy the conditions

$$|a(t, \varphi) - a(t, \varphi_1)| \leq L_1 \|\varphi - \varphi_1\|, \qquad a(t, 0) = b(t, 0) = 0$$
$$|b(t, \varphi) - b(t, \varphi_1)| \leq L \|\varphi - \varphi_1\|, \qquad \varphi, \varphi_1 \in C[-h, 0] \tag{6.2}$$

Examine the stability of the trivial solution of the equation

$$\dot{x}(t) = a(t, x_t) + b(t, x_t)\zeta(t) \tag{6.3}$$

where $\zeta(t)$ is a stochastic process with continuous samples. The definitions of stability for Eq. (6.3) are similar to Definitions 2.1–2.3.

Let the trivial solution of Eq. (6.1) be uniformly exponentially stable. Then there exists the functional $V(t, \varphi)$ satisfying (5.6). Define the function $z(t)$ by conditions $z(t) = x(t)$ for $t \leq s$ and $\dot{z}(t) = a(t, z_t)$ for $t > s$. Then for any $s \geq 0$

$$\dot{V}(s, x_s) = \overline{\lim_{\Delta \to +0}} (1/\Delta)[V(s + \Delta, z_{s+\Delta}) - V(s, x_s)]$$
$$+ \overline{\lim_{\Delta \to +0}} (1/\Delta)[V(s + \Delta, x_{s+\Delta}) - V(s + \Delta, z_{s+\Delta})]$$
$$\leq -(c_4/c_2)V(s, x_s) + c_3|b(s, x_s)\zeta(s)|$$

Hence according to Lemma 1.4.2 from Chapter 1 and (5.5)

$$\|x_t\| \leq (c_2/c_1)\|\varphi\| \exp[-(c_4/c_2)t]$$
$$+ (c_3 L/c_1) \int_0^t \exp[-c_4(t-s)/c_2] \|x_s\| |\zeta(s)| \, ds$$

Applying Lemma 1.4.1 to the function $\gamma(t) = \|x_t\| \exp[(c_4/c_2)t]$ we get

$$\gamma(t) \leq (c_2/c_1)\|\varphi\| \exp\left[(c_3 L/c_1) \int_0^t |\zeta(s)| \, ds\right]$$

Further

$$\|x_t\| \le (c_2/c_1)\|\varphi\| \exp\left[-c_4 t/c_2 + c_3 L/c_1 \int_0^t |\zeta(s)|\, ds\right]$$

Require that

$$E \exp 2\left[-c_4 t/c_2 + c_3 L/c_1 \int_0^t |\zeta(s)|\, ds\right] \le \exp(-\beta t), \qquad \beta \ge 0 \quad (6.4)$$

Then $E\|x_t\|^2 \le (c_2/c_1)^2 E\|\varphi\|^2 \exp(-\beta t)$. Thus the following theorem holds.

Theorem 6.1. *Let condition* (6.4) *be fulfilled and the trivial solution of Eq.* (6.1) *is uniformly exponentially stable. Then the trivial solution of Eq.* (6.2) *is asymptotically mean-square stable.*

6.2. Scalar Equations

Formulate some stability conditions for scalar equations of the form (6.3)

$$\dot{x}(t) = -\int_0^\infty x(t-s)\, dK_0(s) + \beta[t, x(t)]\zeta(t), \qquad t > 0 \quad (6.5)$$

The kernel $k_0(s)$ satisfies the inequality

$$\int_{s_0}^s |dk_0(s)| \le C_1 [\exp(-Cs_0) - \exp(-Cs)], \qquad s \ge s_0 \ge 0 \quad (6.6)$$

Also assume that one of the following conditions is valid:

(1) $$\beta_{00} > 0,\ \alpha_{10} < 1, \quad (6.7)$$

(2) the kernel $k_0(s)$ has a jump at zero of value $a > 0$ and

$$a > \int_{+0}^\infty |dk_0(s)|, \qquad \alpha_{10} < \infty \quad (6.8)$$

(3) $k_0(s)$ is nondecreasing and $k_0(s) = k_0(h)$ for $s \ge h \ge 0$,

$$\int_0^h dk_0(s) > 0, \qquad 2h \int_0^h dk_0(s) < \pi \quad (6.9)$$

The continuous function $\beta(t, x)$ is such that

$$|\beta(t, x_1) - \beta(t, x_2)| \le L|x_1 - x_2|, \qquad \beta(t, 0) = 0$$

Denote $z(t)$ the solution of Eq. (6.5) for $\beta \equiv 0$ with initial conditions $z(0) = 1$, $z(s) = 0$, $s < 0$. By virtue of (6.6) and one of conditions (6.7)–(6.9) there exist constants γ_1, γ_2 such that $|z(t)| \le \gamma_1 \exp(-\gamma_2 t)$, where $0 < \gamma_2 < C$, $\gamma_1 > 0$.

Theorem 6.2. *Let the above formulated assumptions about Eq. (6.5) be valid. Then:*

(a) the trivial solution of Eq. (6.5) is asymptotically stable with respect to probability if $\sup_{t \geq 0} E|\zeta(t)| < \gamma_2(\gamma_1 L)^{-1}$

$$P\left\{\frac{1}{t} \int_0^t [\zeta(s) - E\zeta(s)] \, ds \to 0\right\} = 1, \quad t \to \infty$$

(b) the trivial solution of Eq. (6.5) is asymptotically p-stable for $p \leq \beta_1(2\gamma_1 L)^{-1}$, *if there exists a positive* β_1, β_2 *such that*

$$E \exp\left\{\beta_1 \int_0^t |\zeta(s)| \, ds\right\} \leq \exp(\beta_2 t), \quad \gamma_2 \beta_1 > \gamma_1 L \beta_2$$

Other stability conditions for equations disturbed by the process $\zeta(t)$ can be found in Kolmanovskii and Nosov [108(5)], and Shaichet [199(2)].

§7. NEUTRAL-TYPE STOCHASTIC FUNCTIONAL DIFFERENTIAL EQUATIONS

7.1. Definition of SNFDEs

An SNFDE may be obtained from deterministic Eq. (3.3.2) by adding random perturbations to their right side. This class of equations was introduced in Kolmanivskii [108(5)]. We now present some results for the following equations

$$d[x(t) - G(t, x_t)] = a(t, x_t) \, dt + b(t, x_t) \, d\xi(t), \quad t \geq 0 \quad (7.1)$$

Here $x \in R_n$ and the continuous functional G defined on

$$[0, \infty) \times CB[-\infty, 0]$$

satisfies the conditions

$$|G(t, \varphi) - G(t, \psi)| \leq \int_0^\infty |\varphi(-s) - \psi(-s)| \, dK_1(s) \quad (7.2)$$

$$\int_0^\infty dK_1(s) < 1, \quad \sup_{t \geq 0} |G(t, 0)| < \infty$$

where $K_1(s)$ is a nondecreasing function. The process $\xi(t)$ is a standard Wiener one. The functionals $a(t, x_t)$ and $b(t, x_t)$ are just the same as in Eq. (1.1). The initial conditions for Eq. (7.1) has the form (1.2), (1.5). The solution of problem (7.1), (1.2) is understood in the sense of corresponding integral identity.

7.2. Existence Theorem

Equation (7.1) differs from Eq. (1.1) only by the term $G(t, x_t)$. But it is easy to generalize the theorems about the existence, uniqueness and increase of the solutions of problem (7.1), (1.2) and also the theorems about the existence and properties of stationary solutions known for $G = 0$. We state here without proof the following theorems.

Theorem 7.1. *Let Eq. (7.1) and the initial function satisfy the formulated requirements. Then there exists the unique solution of problem (7.1), (1.2). This solution has a bounded fourth moment on any finite time interval and is measurable relative to the processes $\varphi(\theta)$, $\theta \leq 0$ and $\xi(s)$, $0 \leq s \leq t$. The process x_t is Markovian.*

The proof of Theorem 7.1 is founded on the method of successive approximations [91, 108(5)].

Theorem 7.2. *Let the requirements of Theorem 7.1 be fulfilled and also the functionals a, b, G be independent of t. Assume that there exists a solution of Eq. (7.1) with bounded fourth moments. Then there exists a stationary solution of Eq. (7.1).*

7.3. Stability Conditions

Assume that $G(t, 0) = a(t, 0) = b(t, 0) = 0$. Then there exists a trivial solution of Eq. (7.1). The stability of this solution can be understood in different senses, for example, in the mean-square sense, with probability 1 etc. Here we consider the mean-square stability (see Definition 2.1). Analysis of the stability SNFDE (7.1) as well as the deterministic NFDEs is reasonable to reduce to the considerations to two auxiliary problems. The first one is a construction of a nonnegative functional with a nonpositive derivative. The second one is investigation of the stability of the trivial solution of some auxiliary functional inequality. The form of this inequality depends essentially on the sense in which the stability of Eq. (7.1) is investigated. So for study of mean-square stability it is necessary to consider the inequality

$$E|x(t) - G(t, x_t)|^2 \leq C_0, \qquad t \geq 0 \qquad (7.3)$$

The solution of inequality (7.3) for $t \geq 0$ is determined by initial condition (1.2). The mean-square stability of inequality (7.3) means that for any $\varepsilon > 0$ there exists a $\delta(\varepsilon) > 0$ such that $E|x(t)|^2 < \varepsilon$ if $C_0 < \delta(\varepsilon)$, $E|\varphi(\theta)|^2 < \delta(\varepsilon) \leq 0$.

Theorem 7.3. *Suppose that Eq. (7.1) satisfies the formulated requirements and the trivial solution of inequality (7.3) is mean-square stable. Further, let there*

§7. Neutral-Type Stochastic Functional Differential Equations

exist on $[0, \infty) \times CB[-\infty, 0]$ *a continuous functional* $V(t, \varphi)$ *of the form*

$$V(t, \varphi) = W(t, \varphi) + |\varphi(0) - G(t, \varphi)|^2 \qquad (7.4)$$

such that

$$EV(t, \varphi) \leq C \sup_{\theta \leq 0} E|\varphi(\theta)|^2$$

$$EV(t_2, x_{t_2}) - EV(t_1, x_{t_1}) \leq 0 \qquad (7.5)$$

The functional $W(t, \varphi)$ *in* (7.4) *is defined on* $[0, \infty) \times CB[-\infty, 0]$, *continuous, nonnegative and satisfies estimate* (7.5). *Then the trivial solution of Eq.* (7.1) *is mean-square stable.*

EXAMPLE 7.1. Consider the neutral-type stochastic equation arising in the theory of aeroelasticity (Chapter 3, §2)

$$d\left[x(t) - \int_0^\infty x(t-s)\lambda(s)\,ds\right]$$
$$= \left[-\int_0^\infty x(t-s)\,dK_0(s) + a(t, x_t)\right]dt + b(t, x_t)\,d\xi(t)$$
$$t \geq 0, \qquad x(t) \in R_1$$

Assume that $K_0(s)$ has a jump at zero of the value $b_1 > 0$, the bounded function $\lambda(s)$ is continuously differentiable and

$$\alpha_{01} = \int_0^\infty \lambda(s)\,ds < 1, \qquad \int_0^\infty |\dot\lambda(s)|\,ds < \infty$$

In addition, conditions (1.5) are fulfilled and

$$b_1(1 - \alpha_{01}) - (1 + \alpha_{01})\left[\int_0^\infty |dK_0(s)| + r_1\right] + \frac{1}{2}r_2^2 > 0$$

Then trivial solution of Eq. (7.1) is asymptotically mean-square stable. For the proof consider the functional

$$V(t, x_t) = \left[x(t) - \int_0^\infty x(t-s)\lambda(s)\,ds\right]^2 + \int_0^\infty [(\alpha_{00} + r_1^{-1})|\lambda(s)|\,ds$$
$$+ (\alpha_{01}r_1 + r_1^{-1})\,dR_1(s) + dR_2(s)]\int_{t-s}^t x^2(t_1)\,dt_1 + (1 + \alpha_{01})$$
$$\cdot \int_{+0}^\infty dK_0(s)\int_{t-s}^t x^2(t_1)\,dt_1$$

(for details see Kolmanovskii [105(5)], Kolmanovskii and Nosov [108(5)]).

§8. STABILITY OF LINEAR AUTONOMOUS EQUATIONS

8.1. Systems of Linear Equations

Derive the asymptotic mean-square stability conditions for systems [199(1)].

$$dx(t) = Ax(t)\,dt + \sum_{r=1}^{N} B_r x(t-h)\,d\xi_r(t), \qquad h \geq 0, \quad t \geq 0 \qquad (8.1)$$

where $x \in R_n$ and $\xi_r(t)$ are scalar independent Wiener processes. The elements a_{ij} and b_{ij}^r of matrices A and B_r are constant.

Notice that necessary and sufficient conditions of asymptotic stability of the deterministic equation $\dot{x} = Ax$ are [155]

$$\Delta_i > 0, \qquad i = 1, \ldots, n \qquad (8.2)$$

Here Δ_i are principal diagonal minors of matrix S,

$$S = \begin{vmatrix} -S_1 & -S_3 & -S_5 & \cdots & 0 \\ 1 & S_2 & S_4 & \cdots & 0 \\ 0 & -S_1 & -S_3 & \cdots & 0 \\ \vdots & \vdots & \vdots & & \vdots \\ 0 & 0 & 0 & \cdots & (-1)^n S_n \end{vmatrix}$$

$$S_k = \sum_{1 \leq i_1 < \cdots < i_k \leq n} \begin{bmatrix} a_{i_1 i_1} & \cdots & a_{i_1 i_1} \\ \vdots & & \vdots \\ a_{i_k i_1} & \cdots & a_{i_k i_k} \end{bmatrix}, \qquad k = 1, \ldots, n$$

Denote

$$\omega(x) = \sum_{r=1}^{N} x' B_r' B_r x = \sum_{i,j=1}^{n} d_{ij}(x)$$

$$d_{ij}(x) = \sum_{k,s=1}^{n} \sum_{r=1}^{N} b_{ik}^r b_{js}^r x_s x_k$$

Lemma 8.1. *The trivial solution of Eq. (8.1) is asymptotically mean-square stable if there exists a positive-definite quadratic form* $V(x) = \sum_{i,j=1}^{n} V_{ij} x_i x_j = x'Vx$ *such that* $x'(AV + VA)x = -\omega(x)$ *and also the quadratic form* $\alpha(x) = \sum_{i,j=1}^{n} d_{ij}(x)(1 - V_{ij})$ *is positive-definite.*

Proof. Let L be the generating operator of Eq. (8.1). Consider the functional

$$W(t, x_t) = V[x(t)] + \int_{t-h}^{t} x'(s) \sum_{r=1}^{N} B_r' V B_r x(s)\,ds \qquad (8.3)$$

§8. Stability of Linear Autonomous Equations

Then from Itô's formula

$$LW(t, x_t) = -\alpha(x) \tag{8.4}$$

Hence the validity of Lemma 8.1 follows from (8.3), (8.4) and Theorem 2.1.

REMARK. Let $d_{ij} = 0$ for $i \neq j$. Then $\alpha(x)$ is positive-definite if all $V_{ii} < 1$.

Find V_{ij} and $\alpha(x)$. Let $\Delta_{1,m}$ be the cofactor of the first row and m th column of the determinant Δ_n. Determine the numbers q_{ij}^r by the equation

$$(-1)^n \sum_{k,s=1}^{n} \omega_{ks} D_{ik}(\lambda) D_{js}(-\lambda) = \sum_{r=0}^{n-1} q_{ij}^r \lambda^{2(n-r-1)}$$

Here ω_{ks} are coefficients of quadratic form $\omega(x)$ and $D_{ik}(\lambda)$ are cofactors of the determinant $D(\lambda)$

$$D(\lambda) = \begin{vmatrix} a_{11} - \lambda & \cdots & a_{1n} \\ \vdots & & \vdots \\ a_{n1} & \cdots & a_{nn} - \lambda \end{vmatrix}$$

Then

$$V_{ij} = \Delta_{ij}/2\Delta_n, \qquad \Delta_{ij} = \sum_{m=0}^{n-1} q_{ij}^m \Delta_{1,m+1} \tag{8.5}$$

$$\alpha(x) = \sum_{k,s=1}^{n} \alpha_{ks} x_k x_s, \qquad \alpha_{ks} = \sum_{i,j=1}^{n} \sum_{r=1}^{N} b_{ik}^r b_{js}^r (1 - V_{ij})$$

Denote by δ_i principal diagonal minors of the matrix (α_{ks}).

Theorem 8.1. *Let conditions (8.2) be fulfilled and $\delta_i > 0$, $i = 1, \ldots, n$. Then the trivial solution of system (8.1) is asymptotically mean-square stable.*

The proof follows from Lemma 8.1 and relation (8.5).

EXAMPLE 8.1. Given two equations

$$dx_i(t) = \sum_{k=1}^{2} a_{ik} x_k(t) \, dt + b_i x_i(t-h) \, d\xi_i(t), \qquad i = 1, 2$$

Sufficient conditions for asymptotic mean-square stability of these equations are

$$S_1 = a_{11} + a_{22} < 0, \qquad S_2 = a_{11}a_{22} - a_{12}a_{21} > 0$$
$$\max[b_1^2(S_2 + a_{22}^2), b_2^2(S_2 + a_{11}^2)] < -2 \cdot S_1 \cdot S_2$$

8.2. Corollary for nth-Order Equations

Consider a linear SRFDE

$$x^{(n)}(t) + \sum_{i=1}^{n}\left[a_i x^{(n-i)}(t) + \sum_{j=1}^{N} \beta_{ij} x^{(n-i)}(t-h)\dot{\xi}_j(t)\right] = 0 \tag{8.6}$$

This equation is equivalent to system of the form (8.1)

$$dx_1(t) = x_2(t)\, dt, \ldots, dx_{n-1}(t) = x_n(t)\, dt$$

$$dx_n = -\sum_{i=1}^{n} a_i x_{n-i+1}(t)\, dt - \sum_{i=1}^{n}\sum_{j=1}^{N} \beta_{ij} x_{n-i+1}(t-h)\, d\xi_j(t) \tag{8.7}$$

For system (8.7) matrices A and S are

$$A = \begin{vmatrix} 0 & 1 & 0 & \cdots & 0 \\ 0 & 0 & 1 & \cdots & 0 \\ \vdots & \vdots & \vdots & & \vdots \\ -a_n & -a_{n-1} & -a_{n-2} & \cdots & -a_1 \end{vmatrix}, \quad S = \begin{vmatrix} a_1 & a_3 & \cdots & 0 \\ 1 & a_2 & \cdots & 0 \\ 0 & a_1 & \cdots & 0 \\ \vdots & \vdots & & \vdots \\ 0 & 0 & \cdots & a_n \end{vmatrix} \tag{8.8}$$

Functions $\omega(w)$ and $d_{ij}(x)$ are $d_{ij}(x) = 0$ if $i + j < 2n$, $\omega(x) = d_{nn}(x)$

$$d_{nn}(x) = \sum_{k,s=1}^{n} \sum_{j=1}^{N} \beta_{kj}\beta_{sj} x_{n+1-s} x_{n+1-k} \tag{8.9}$$

Lemma 8.2. *Trivial solution of Eq. (8.6) is asymptotically mean-square stable if there exists a positive definite quadratic form $V(x) = x'Vx = \sum_{i,j=1}^{n} V_{ij} x_i' x_j$ such that*

$$x'(A'V + VA)x = -d_{nn}(x), \quad V_{nn} < 1$$

Theorem 8.2. *The trivial solution of Eq. (8.6) is asymptotically mean-square stable if*

$$\Delta_i > 0,\ (i = 1, \ldots, n), \quad \Delta_n > \Delta/2$$

Here Δ_i are principal diagonal minors of matrix S in (8.8) and

$$\Delta = \begin{vmatrix} q_{nn}^{(0)} & q_{nn}^{(1)} & \cdots & q_{nn}^{(n-1)} \\ 1 & a_2 & \cdots & 0 \\ \vdots & \vdots & & \vdots \\ 0 & 0 & \cdots & a_n \end{vmatrix}$$

$$q_{nn}^{(r)} = \sum_{p+q=2(r+1)} (-1)^{q+1} \sum_{j=1}^{N} \beta_{pj}\beta_{qj}, \quad r = 0, \ldots, n-1$$

§8. Stability of Linear Autonomous Equations

8.3. Necessary and Sufficient Stability Conditions of Scalar Equations

Consider the scalar equation

$$dx(t) = \int_{-h}^{0} x(t+s)\, dK(s)\, dt \\ + \int_{-h}^{0} x(t+s)\, dR(s)\, d\xi(t), \qquad t \geq 0 \tag{8.10}$$

with initial condition (1.2). Here $K(s)$ and $R(s)$ are functions with bounded variations on the interval $0 \leq s \leq h$. The necessary and sufficient conditions

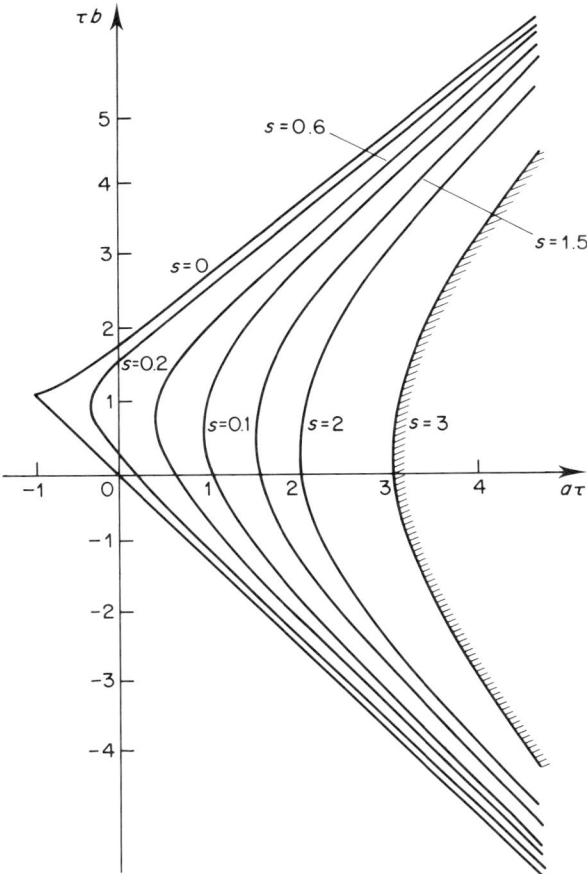

Fig. 4.13. Stability domains for Eq. (8.13) for $s \in [0, 3]$.

for asymptotic mean-square stability of the trivial solution of the system (8.10) are [230(2)]:

(1) all roots of the characteristic equation

$$z - \int_{-h}^{0} \exp(zs)\, dK(s) = 0 \qquad (8.11)$$

lie in the left half plane;

(2) $\quad \dfrac{1}{\pi} \displaystyle\int_{0}^{\infty} \left| \int_{-h}^{0} \exp(is\theta)\, dR(\theta) \right|^{2} \left| i\theta - \int_{-h}^{0} \exp(ist)\, dK(t) \right|^{-2} ds < 1 \quad (8.12)$

Remark that from condition (1) it follows that deterministic system (8.10) with $R = 0$ is asymptotically stable. Let, for example, Eq. (8.10) have the form

$$dx(t) = [-ax(t) - bx(t-\tau)]\, dt + cx(t-\tau)\, d\xi(t) \qquad (8.13)$$

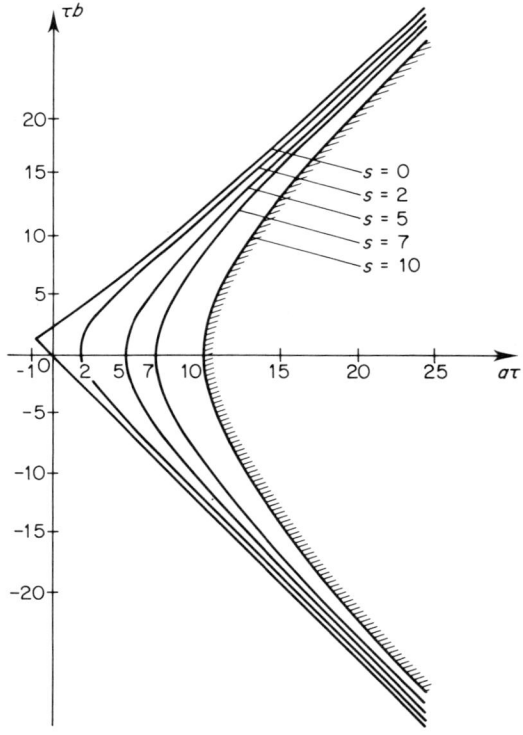

Fig. 4.14. Stability domains for Eq. (8.13) for $s \leq 10$.

§8. Stability of Linear Autonomous Equations

From (8.11), (8.12) we obtain that system (8.13) is stable in the case $|b| \geq |a|$ if

$$c^2 < 2(1 + bk^{-1} \sin k\tau)^{-1}(a + b \cos kj), \qquad k = \sqrt{b^2 - a^2} \qquad (8.14)$$

and in the case $|b| \leq |a|$ if

$$c^2 < 2(1 + bk^{-1} shk\tau)^{-1}(a + bchk\tau), \qquad k = \sqrt{a^2 - b^2} \qquad (8.15)$$

The stability domains of system (8.13) given by inequalities (8.14), (8.15) are represented in Figs. 4.14 and 4.13 for different values of the parameter $s = c^2\tau/2$. The domain of stability is to the right of the boundary.

References

1. Achmerov, R. R., Kamenskii, M. I., Potapov, A. S., Rodkina, A. E. and Sadovskii, B. N., Theory of equations of neutral type. *Naucn.Tehn. Informacija (VINITI)*, **16** (1982).
2. Afanasjev, V. N. and Rodionov, A. M., Stabilization of a class of nonlinear systems with delay. *Differencial'nye Uravnenija* **17** (5), 914–916 (1981).
3. Alekseevskaja, N. L. and Gromova, P. S., "Liapunov Direct Method for Retarded Differential Equations—Differential Equations with Deviating Argument," pp. 17–34. Nauka Dumka, 1977.
4. Andronov, A. A. and Mayer, A. G., The simplest linear systems with delay. *Automat. Remote Control* **7** (2–3), 95–106 (1946).
5. Angelov, V. and Bainov, D. D., Existence and uniqueness of the global solution of the initial value problem for neutral type differential-functional equations in Banach space. *Nonlinear Anal.* **3**(6) (1979).
6. Arutunian, N. H. and Kolmanovskii, V. B., "Creep Theory of Nonhomogeneous Bodies", Nauka, Moscow, 1983.
7. Astapov, I. S., Belotzerkovskii, S. M., Kachanov, B. O. and Kochetkov, Ju. A., Systems of integro-differential equations describing unsteady motion of bodies in continuous media. *Differencial'nye Uravnenija* **18**(9), 1628–1636 (1982).
8. Azbelev, N. B. and Sulavko, T. S., About stability of solutions of differential equations with delay. *Differencial'nye Uravnenija* **10**(12), 2091–2100 (1974).
9. Azjan, Ju. M. and Migulin, V. V., About auto-oscillation in systems with delayed feedback. *Radiotek. Elektron.* **1**(4), 418–430 (1956).
10. Babskii, V. G. and Myshkis, A. D., Mathematic models in biology, connected with regard of delays (Appendix) Murray, J. D., "Lectures on nonlinear-differential equations," 383–394. *Models in Biology*, MIR, Moscow, 1983.
11. Bailey, H. R. and Reeve, E. B., Mathematical models describing the distribution of I^{131}-albumin in man. *J. Lab. Clin. Med.* (60), 923–943 (1962).
12. Bakli, P. S. Control of the processes with delay. *Trudy IFAK, Moscow*, 95–111 (1961).
13. Banach, S. "Théorie des Opérations Lineaires." Warszawa, Poland, 1932.
14. Banks, J. T. (1) Representations for solutions of linear functional differential equations. *J. Differ. Eq.* **5**(2), 399–409 (1969). (2) "Modeling and Control in the Biomedical Sciences," *Lecture Notes in Biomath.* **6** (1975). (3) Parameter identification techniques for physiological control systems. *In* "Mathematical Aspects of Physiology," *Lectures in Appl. Math.* **19**, 361–385 (1981).
15. Barbashin, E. A., "Introduction to Stability Theory." Nauka, Moscow, 1967.
16. Barnea, D. I., A method and new results for stability and instability of autonomous functional differential equations. *SIAM J. Appl. Math.* (17), 681–697 (1969).
17. Bellman, R., On the existence and boundedness of solutions of nonlinear differential difference equations. *Ann. of Math.* **50**(2), 347–355 (1949).
18. Bellman, R. and Cooke, K., "Differential-Difference Equations," Academic Press, New York, 1963.

19. Belotzerkiovskii, S. M., Kochetkov, Ju. A., Krasovskii, A. A. and Novitzkii, V. V., "Introduction to aeroautoelasticity." Nauka. Moscow, 1980.
20. Belotzerkovskii, S. M., Skripach, B. K. and Tabachnikov, V. G., "Wing in unsteady gas flow." Nauka, Moscow, 1971.
21. Belykh, L. N. and Marchuk, G. I., Chronic forms of a disease and their treatment according to mathematical immune response models. *Lecture Notes in Control and Information Sci.* **18**, 79–86 (1979).
22. Bodner, V. A., "Operator and flying vehicle." Mashinostroenie, Moscow, 1976.
23. Bodner, V. A., Rjazanov, Ju. A. and Shaimardanov, F. A., "Systems of control of aircraft engines." Mashinostroenie, Moscow, 1973.
24. Bodner, V. A., Zakirov, R. A. and Smirnova, I. I., "Aircraft trainers." Mashinostroenie, Moscow, 1978.
25. Bogomolov, V. L., Control of power of hydrostation. *Automat. Remote Control* (4-5), 103–129 (1941).
26. Braddock, R. D. and Driessche, P. van den, Population models with time delay. *Math. Sci.* **5**(1), 55–66 (1980).
27. Brauer, F., Stability of some population models with delay. *Math. Biosci.* **33**(2), 345–358 (1977).
28. Brayton, R., Nonlinear oscillations in a distributed network. *Quart. Appl. Math.* **24**, 289–301 (1976).
29. Brewer, D. W., The asymptotic stability of a nonlinear functional differential equation of infinite delay. *Houston J. Math.* **6**(3), 321–330 (1980).
30. Burton, T. A., "Volterra Integral and Difference Equations," Academic Press, New York, 1983.
31. Butkovskii, A. G., Structural method for system with distributed parameters. *Automat. Remote Control*, No. 5 (1975).
32. Callender, A. and Stevenson, A. G., Time lag in a control system. *Proc. Soc. Chem. Ind.* **18**(1), 108–117 (1939).
33. Cargill, T. F. and Meyer, R. A., Wages, prices and unemployment: distributed lag estimates. *J. Amer. Statist. Assoc.* **69**(345), 98–107 (1974).
34. Castelan, W. B., A Liapunov functional for a matrix retarded difference-differential equation with several delays. *Lecture Notes in Math.* (799), 82–118 (1980).
35. Castelan, W. B. and Infante, E. F., A Liapunov functional for a matrix neutral difference-differential equation with one delay. *J. Math. Anal. Appl.* **71**(1), 105–130 (1979).
36. Cesari, L., "Asymptotic Behavior and Stability Problems in Ordinary Differential Equations," Springer, New York, 1959.
37. Chebotarev, N. G. and Meiman, N. N., Problem of Routh-Gurwitz for polynomials and entire functions. *Trudy Math. Inst. Steklov* **26** (1949).
38. Chernousko, F. L. and Kolmanovskii, V. B. (1) "Optimal Control under Stochastic Disturbances." Nauka, Moscow, 1978. (2) Computational and approximate methods of optimal control *J. Soviet. Math.* **12**, 310–353 (1979).
39. Chetaev, N. G., "Stability of Motion." Nauka, Moscow, 1965.
40. Chipot, M., On the equations of age-dependent population dynamics. *Arch. Rational Mechanics Anal.* **82**(1), 11–25 (1983).
41. Chistjakov, P. G., "Accuracy of control system of JRD and TRD." Mashinostroenie, Moscow, 1977.
42. Coleman, B. D., A general theory of dissipation in materials with memory. *Arch. Rational Mech. Anal.* 255–274 (1968).
43. Coleman, B. D. and Dill, H., On the stability of certain motions of incompressible materials with memory. *Arch. Rational Mech. Anal.* **30**, 197–224 (1968).

44. Coleman, B. D. and Owen, D. R., On the initial-value problem for a class of functional differential equations. *Arch. Rational Mech. Anal.* **55**, 275–299 (1974).
45. Cooke, K. L. and Yorke, J. A., Equations modelling population growth, economic growth and gonorrhea epidemiology. *In* "Ordinary Differential Equations" (L. Weiss, ed.), Academic Press, New York, 1972, pp. 35–55.
46. Corduneanu, C. (1) Integral Equations and Stability of Feedback Systems. Academic Press, New York, 1973. (2) Stability problems for some feedback systems with delay. *In* "Modern Trends in Cybernetics and Systems." Vol. 2. pp. 321–328, Springer-Verlag, Berlin and New York, 1977.
47. Corduneanu, C. and Lakshmikantham, V., Equations with unbounded delay: a survey. *Nonlinear Anal. Theory Methods Appl.* **4**(5), 831–878 (1980).
48. Cruz, M. A. and Hale, J. K., Stability of functional differential equations of neutral type. *J. Differ. Eq.* **7**(2), 334–355 (1970).
49. Cryer, C. W., Numerical methods for functional differential equations. In "Delay and Functional Differential Equations and Their Applications," pp. 17–101, Academic Press, New York, 1972.
50. Cushing, C. M., "Integrodifferential Equations and Delay Models in Population Dynamics." *Lecture Notes in Biomath.* **20** (1977).
51. Danilushkin. A. I. and Rapoport, E. Ja., Algorythm of process functioning of continuously-successive inductive heating. *In* "Algorythmizing and automatization of technological processes in industrial mountings" Vol. 7. Kuibyschev, 1976.
52. Datko, R., The uniform asymptotic stability of certain neutral differential-difference equations. *J. Math. Anal. Appl.* **58**(3), 510–526 (1977).
53. Differential Equations with Deviating Argument. Nauka Dumka, Kiev, 1977.
54. Delay and Functional Differential Equations and Their Applications. (K. Schmitt, ed.) Academic Press, New York, 1972.
55. Dibrov, B. F., Livshits, M. A. and Volkenstein, M. V. The effect of a time lag in the immune reaction. *Lecture Notes in Control and Information Sci.* **18**, 87–94 (1979).
56. Draljuk, B. N. and Sinaiskii, G. V., "Systems of Control for Plants with Transport Delays." Energia Moscow, (1969).
57. Driver, R. D. (1) A functional differential system of neutral type arising in a two-body problem of classical electrodynamics. *In* "Nonlinear Differential Equations and Nonlinear Mechanics" pp. 474–484. Academic Press, New York, 1963. (2) "Ordinary and Delay Differential Equations" Springer-Verlag, Berlin and New York, 1977.
58. El'sgol'tz, L. E. (1) Stability of solutions of differential-difference equations. *Uspehi Mat. Nauk*, **9**(4), 95–112 (1954). (2) To the influence on stability of small deviation of argument. *Trudy Sem. Teor. Differencial. Uravnenii s Otklon. Argumentom Univ. Druzby Narodov Patrisa Lumumby* **1**, 114–115, (1962). (3) Qualitative methods in mathematical Analysis. *Trans. Math. Monographs* **12**, (1964).
59. El'sgol'tz, L. E. and Norkin, S. B., "Introduction to the Theory and Application of Differential Equations with Deviating Arguments," Academic Press, New York, 1973.
60. Faerman, E. Ju., "Problem of Long-Range Planning." Nauka, Moscow, 1971.
61. Fichtengol'tz, G. M., "Course of Differential and Integral Calculus," Vol. 3. Fizmatgiz, Moscow, 1960.
62. Gerasimov, V. G., "Theoretical Foundation of Control of Heat Processes." Gosenergoizdat, Moscow, 1949.
63. Germaidze, V. E. and Krasovskii, N. N., Stability under steady acting disturbances. *Appl. Math. Mech.* **21**(6), 769–774 (1957).
64. Gihman, I. I. and Skorohod, A. V., "Stochastic Differential Equations," *Ergeb. Math. Grenzgeb.* **72** (1972).

65. Ginzburg, P. E., Application of Liapunov direct method for investigation of oscillations in linear systems with delay. *Differencial'nye Uravnenija* **7**(10), 1903–1905 (1971).
66. Gonorovskii, I. S., To the problem of auto-oscillations in a high frequency generator with delayed feedback. *Radiotehnika (Kharkov)* **14**(1), 25–33 (1959).
67. Gorbunov, V. P. and Shichov, S. B., "Nonlinear Dynamics of Nuclear Reactors." Atomizdat, Moscow, 1975.
68. Górecki, H., "Analysis and synthesis of controlled systems with delay." Mashinostroenie, Moscow, 1974.
69. Gorelik, G., To the theory of feedback with delay, *J. Tech. Phys.* **9**(5), 450–454 (1939).
70. Gorjachenko, V. D., "Methods of Stability Investigations of Nuclear Reactors." Atomizdat, Moscow, 1977.
71. Gorjachenko, V. D. and Ivanov, B. N., Dynamics of interaction of two populations as objects with delay. In: "Dynamics of Biological Systems," pp. 9–25. Gorjkii, 1978.
72. Gromova, P. S. (1) Investigation of stability of linear differential-difference equations of neutral type in critical cases. *Nonlinear Vibration Problem* **4**, 149–158 (1973). (2) Stability of solutions of linear differential-difference equations of n-th order. *Trudy Sem. Teor. Differencial. Uravnenii s Otklon. Argumentom Univ. Druzby. Narodov Patrisa Lumumby* **9**, 31–39 (1975).
73. Gromova, P. S. and Markos, L. P., Method of Liapunov vector functions for retarded systems. *Trudy Sem. Teor. Differencial. Uravnenii s Otklon. Argumentom Univ. Druzby. Nardov Patrisa Lumumby*, 14–22 (1979).
74. Grossman, S. E. and Yorke, J. A., Asymptotic behavior and exponential stability criteria for differential delay equations. *J. Differential Equations* **12**(2), 236–255 (1972).
75. Gumowski, I., Sensitivity of certain dynamic systems with respect to a small delay. *Automatica—J. IFAC* **10**(6), 659–674 (1974).
76. Gurtin, E. M. (1) Thermodynamics and the energy criterion for stability. *Arch. Rational Mech. Anal.* **52**(2), 93–103 (1973). (2) Some questions and open problems in continuum mechanics and population dynamics. *J. Differential Equations* **48**(2), 293–312 (1983).
77. Hadeler, K. P., "Delay Equations in Biology." *Lecture Notes in Math.* **730**, 136–156 (1979).
78. Hahn, W. (1) Uber Differential-Differenzengleichungen mit anomalen Losungen. *Math. Ann.* **133**, (1957). (2) "Stability of Motion," Springer-Verlag, New York and Berlin, 1967.
79. Halanay, A. (1) Perron's condition in the theory of general systems with after-effect. *Mathematica (Cluj)* **2**(2), 257–267 (1960). (2) Stability theory of linear periodic systems with delay. *Rev. Math. Pures Appl.* **6**(4), 633–653 (1961). (3) Systèmes à retard. Résultats et problèmes. *Tagungsber. ICNO-III (Berlin)* (1965). (4) "Differential Equations: Stability, Oscillations, Time Lags." Academic Press, New York, 1966.
80. Halanay, A. and Yorke, J. A., Some new results and problems in the theory of differential-delay equations. *SIAM Rev.* **13**, 55–80 (1971).
81. Hale, J. K. (1) A stability theorem for functional-differential equations. *Proc. Nat. Acad. Sci. USA*, **50**(5), 942–946 (1963). (2) Asymptotic behavior of the solutions of differential-difference equations. *Trudy ICNO-1 (Kiev)* **2**, 409–426 (1963). (3) Sufficient conditions for stability and instability of autonomous functional-differential equations. *J. Differential Equations* **1**, 452–482 (1965). (4) "Theory of functional differential equations," Springer-Verlag, New York and Berlin, 1977.
82. Hale, J. K. and Ize, A. F., On the uniform asymptotic stability of functional differential equations of the neutral type. *Proc. Amer. Math. Soc.* **28**(1), 100–106 (1971).
83. Hale, J. K. and Meyer, K. R., A class of functional equations of neutral type. *Mem. Amer. Math. Soc.* (76), 1–65 (1967).
84. Has'minskii, R. Z., "Stochastic stability of differential equations." Noordhofof, Sijthoff, 1980.

85. Hausrath, A., Stability in the critical case of purely imaginery roots for neutral functional differential equations. *J. Differential Equations* (13), 329–357 (1973).
86. Hino, Y., On stability of some functional differential equations. *Funkcial. Ekvac.* **14**, 47–60 (1971).
87. Husainov, D. Ja. and Sharkovskii, A. N., Stability of solutions of differential equations with delay. *In* "Functional and Differential-Difference Equations," pp. 141–147. Inst. of Math. of Ukr. Acad. Sci., Kiev, 1974.
88. Hutchinson,. G. E., Circular causal systems in ecology. *Ann. N.Y. Acad. Sci.* **50**, 221–246 (1948).
89. Infante, E. F. and Castelan, W. B., A Liapunov functional for a matrix difference-differential equation. *J. Differential Equations*, **29**(3), 439–451 (1978).
90. Infante, E. F. and Slemrod, M., Asymptotic stability criteria for linear systems of differential difference equations of neutral type and their discrete analogues. *J. Math. Anal. Appl.* (38), 399–415 (1972).
91. Itô, K. and Nisio, M., On stationary solutions of a stochastic differential equation. *J. Math. Kyoto Univ.* **4**, 1–75 (1964).
92. Ize, A. F. and dos Reis, J. G., Stability of perturbed neutral functional differential equations. *Nonlinear Anal.* **2**(5), 563–571 (1978).
93. Kabakov, I. P., Concerning the control process for the steam process for the steam pressure. *Inžener. Sbornik* (2), 27–60, 61–76 (1946).
94. Kabal'nov, Ju. S., Homjakov, I. M. and Il'jasov, B. G., To the method of control of plants with pure delay. *Izv. Vusov. Electromechanica* (8), 863–866 (1978).
95. Kamenskii, G. A. (1) To the general theory of equations with deviating argument. *Dokl. AN USSR* **120**(4), 697–700 (1958) (2) Existence, uniqueness and continuous dependence from initial data of solutions of systems of differential equations with deviating argument of neutral type. *Matem. Sborn.* **55**(4) (1961).
96. Kantorovich, L. V. and Akilov, G. P., "Functional Analysis." Nauka, Moscow, 1977.
97. Kaplan, J. L. and Yorke, J. A., On the stability of a periodic solution of a differential delay equation. *SIAM J. Math. Anal.* **6**(2), 268–282 (1975).
98. Kato, J. (1) On Liapunov-Razumikhin type theorems for functional differential equations. *Funkcial Ekvac.* **16**(3), 225–239 (1973). (2) Stability problem in functional differential equations with infinite delay. *Funkcial. Ekvac.* **21**, 63–80 (1978).
99. Katz, I. Ia., To the stability in the first approximation of systems with stochastic delay. *Prikl. Mat. Meh.* **31**(3), 447–452 (1967).
100. Kir'janen, A. I., Stability of solutions of differential equations with deviating argument. *Lecture Notes.* Leningrad University, 1983.
101. Klimushev, A. I., About asymptotic stability of retarded systems, containing derivatives with small parameters. *Prikl. Mat. Meh.* **26**(1), 52–61 (1962).
102. Kobrinskii, N. E. and Kus'min, V. I., "Accuracy of economic-mathematical models." Finansy and Statistica, Moscow, 1981.
103. Kolesov, Ju. S., To the stability of linear differential-difference equations of neutral type. *Sibirian Math. J.* **20**(2), 317–321 (1979).
104. Kolesov, Ju. S. and Shvitra, D. I., "Auto-oscillation in systems with delay." Mokslas, Vilnuis, 1979.
105. Kolmanovskii, V. B. (1) Stationary solutions of equations with delay. *Problems Inform. Transmission* **3**(1), 50–57 (1967). (2) Application of the Liapunov method to linear systems with lag. *J. Appl. Math. Mech.* **31**, 976–980 (1967). (3) Boundedness of solutions of integro-differential equations with deviating argument. *Trudy Sem. Teor. Differencial. Uravnenii s Otklon Argumentom Univ. Druzby Narodov Patrisa Lumumby* **6**, 207–212 (1968). (4) Ergodic properties of stochastic differential equation solutions. *Theory Probab. Appl.* **14**(1), 142–148

(1969). (5) On the stability of stochastic systems with delay. *Problems Inform. Transmission* **5**(4), 59–67 (1969). (6) On the stability of neutral-type linear equations. Differencial'nye Uravnenija **6**(7), 1235–1246 (1970). (7) The stability of nonlinear systems with lag. *Math. Notes* **7**, 446–450 (1970). (8) Solutions of constant signs of neutral-type equations. *J. Differential Equations* **7**(6), 1116–1119 (1971). (9) Representation of solutions of neutral type equations. *Ukrain. Mat. Z.* **24**(2), 171–178 (1972). (10) Filtration of stochastic processes with delay. *Automat. Remote Control* (1), 42–49 (1974). (11) On the stability of some stochastic differential equations with retarded argument. *Theory Probab. Math. Statist.* (2), 110–120 (1974). (12) On the equalities for second moments of solutions of stochastic differential equations with delay. *Ukrainian Math. J.* **27**(1), 94–97 (1975). (13) Optimal control of stochastic systems with after-effect. (A survey). *Trudy Fourth All-Union Conf. Theory Appl. Differ. Eq. Deviat. Argument*, pp. 177–185. Nauka Dumka, Kiev, 1977.
106. Kolmanovskii, V. B. and Has'minskii, R. Z., Stability of linear systems with retardation. *Izv. Vyss. Ucebn. Zaved. Matematika* **4**(53), 58–65 (1966).
107. Kolmanovskii, V. B. and Maizenberg, T. L., Optimal estimation of system states and problems of control of systems with delay. *Prikl. Math. Meh.* **41**(3), 446–456 (1977).
108. Kolmanovskii, V. B. and Nosov, V. R. (1) On the stability of nonlinear oscillatory systems described by neutral-type equations. *Proc. 5th Internat. Conf. Nonlinear Osc.* **2**, Kiev (1969). *Issue Inst. Math. Vkr. Acad. Sci.* 228–232 (1970). (2) On the stability of first-order neutral type equations. *Prikl. Math. Meh.* **34**, (4) 587–594. (3) Stability of stationary systems with aftereffect, *Avtomat. i Telemeh.* (1), 9–18 (1979. (4) Stability of systems with deviating argument of neutral type. *Prikl. Math. Meh.* **43**(2), 209–218 (1979. (5) "Stability and periodic modes of control systems with aftereffect." Nauka, Moscow, 1981. (6) Stability of neutral-type functional differential equations (a survey) *Nonlinear Anal. Theory Methods Appl.* **6**(9), 873–910 (1982). (7) Asymptotic behavior of neutral type equations. *In* "Oscillation and Solutions Stability of Functional-Differential Equations," pp. 44–53. Institute of Math. AN USSR, Kiev, 1982. (8) On instability of systems with after-effects. *Automat. Remote Control* **44**, (1), part 1, 24–32 (1983) (9) "Stochastic stability and control." (Lecture notes). Institute of Electronic Machines, Moscow, 1983. (10) Retarded systems of neutral type. (a survey) *Automat. Remote Control* **45**(1), 5–36 (1984).
109. Kolmanovskii, V. B. and Shaichet, L. E., Approximate feedback control of quasilinear stochastic retarded systems. *Prikl. Math. Meh.* **42**(6), 978–988 (1978).
110. Kostitzin, V. A., "Symbiose, Parasitisme et Evolution (Etude Mathematique)," Hermann, Paris, 1934.
111. Kozjakin, V. S. and Krasnoselskii, M. A., To the influence of small delays on dynamic of nonlinear systems. *Automat. Remote Control.* (1), 5–8 (1979).
112. Krasnoselskii, M. A., "Shift operator for differential equations." Nauka, Moscow, 1966.
113. Krasnoselskii, M. A., Burd, V. Sh. and Kolesov, Ju. S., "Nonlinear Almost-Periodic Oscillations." Nauka, Moscow, 1970.
114. Krasovskii, N. N. (1) Inversion of theorems of Liapunov direct method and stability in first approximation. *Prikl. Math. Meh.* **20**(2), 255–265 (1956). (2) On the Application of direct Liapunov method to equations with time lag. *Prikl. Math. Meh.* **20**(3), 315–327 (1956). (3) On the asymptotic stability of systems with aftereffect. *Prikl. Math. Meh.* **20**(3), 513–518 (1956). (4) Stability under large initial disturbances. *Prikl. Math. Meh.* **21**(3), 309–319 (1957). (5) "Stability of motion: Applications of Lyapunov's Second Method to Differential Systems and Equations with Delay." Stanford Univ. Press, Stanford, California, 1963. (6) Analytical design of optimal controller in the system with delay. *Prikl. Math. Meh.* **28**(1), 39–51 (1964).
115. Krupnova, N. I. and Shimanov, S. N., Test of stability of linear systems with periodic coefficients and time-lag. *Prikl. Math. Meh.* **36**(3), 533–536 (1972).

116. Kulesko, N. A., About completeness of Floquet solutions of neutral type equations. *Mat. Zametki* **3**(3), 297–306 (1968).
117. Kurbatov, V. G., Stability of functional-differential equations. *Differen. Uravn.* (6), 963–972 (1981).
118. Kwong, R. H., A stability theory for the linear-quadratic-Gaussian problem for systems with delays in the state, control and observations. *SIAM J. Control Optim.* **18**(1), 49–75 (1980).
119. Lakshmikantham, V. and Leela, S. (1) Differential and Integral Inequalities. Academic Press, New York, 1969. (2) Global Results and Stability of Motion. Univ. Rhode Island, Tech. Rep. No. 8, 1970. (3) A technique in stability theory of delay-differential equations. *Nonlinear Anal.* **3**(3), 317–323 (1979).
120. Lakshmikantham, V. and Rao M. R. H., Integro-Differential Equations and Extension of Liapunov's Method. Univ. Rhode Island, Tech. Rep. No. 7, 1969.
121. Lambourne, N. C., Control-surface buzz *Rep. Memoranda* (3364), 1–17 (1962).
122. Landau, I. D., A survey of model reference adaptive techniques: theory and applications. *Automatica—J. IFAC* **10**(4), 353–379 (1974).
123. Lavrentjev, M. A. and Shabat, B. V., "Methods of Theory of Functions of Complex Variabele." Fizmatgiz, Moscow, 1958.
124. Lekus, V. F. and Rovinskii, V. E., "Estimate of Stability of Systems with Delay." Energoatomizdat, Leningrad, 1982.
125. Leontjev, A. F., Differential-difference equations. *Mat. Sb.* **24**(3), 347–374 (1949).
126. Lewis, R. M. and Anderson, B. D. O., Necessary and sufficient conditions for delay-independent stability of linear autonomous system. *IEEE Trans. Automat. Control* **AC-25**(4), 735–739 (1980).
127. Liapunov, A. M., "Stability of Motion." Gostechizdat, Moscow, 1950.
128. Liberman, L. X., Stability of differential-operator equations with delay under disturbances bounded in the mean. *Siberian Math. J.* **4**(1), 138–144 (1963).
129. Lidskii, E. A., Stability of systems with stochastic delay. *Differencial'nye Uravnenija* **1**(1), 96–101 (1965).
130. Lihtarnikov, A. L. and Jakubovich, V. A., Absolute stability of nonlinear systems. Appendix to Russian translation of "Absolute Stability of Retarded Control Systems" pp 287–357. Rasvan Vb.
131. Loginov, G. V. and Oreshina, S. T., Analysis of retarded systems with predictor by methods of root trajectories. *Izv. Vyss. Ucebn. Zaved. Elektromehanika* (11), 1158–1163 (1978).
132. Lubfer. D. E. and Oglesby, W., Applying dead-time compensation for linear predictor process control. *J. Systems Anal. ISA Journal* **8**(11). 53–57 (1961).
133. Lunt, S. T. and Tolman, G., Some effects of delays in control systems. *Meas. Control* **13**(5), 184–185 (1980).
134. Lutzkiv, N. M. (1) To the synthesis of control systems with compensation of delay. *Izv. Vyss. Ucebn. Zaved. Elektromehanika* (8), 889–901 (1972). (2) Systems with compensation of delay and divisor. *Izv. Vyss. Ucebn. Zaved. Elektromehanika* (9), 1003–1007 (1976).
135. MacDonald, N. "Time Lags in Biological Models." *In Lecture Notes Biomath.* (1978).
136. Mahin, V. A., Prisnjakov, V. F. and Belik, N. P., "Dynamics of liquid reactive engine." Mashinostroenie, Moscow, 1969.
137. Maizenberg, T. L., Dirichlet problem for integro-differential equations. *Izv. AN USSR, Ser. Math.* **33**(3), 570–590 (1969).
138. Malkin, I. G., "Stability of Motion." Nauka, Moscow, 1966.
139. Marchuk, G. I. (1) Some mathematical models in immunology. *Proc. 8th IFIP Conf. Opt. Tech.* Heidelberg, 41–62 (1978). (2) "Mathematical Models in Immunology and Their Interpretation." *In Lecture Notes in Control and Information Sci.* **18**, 114–129 (1979). (3) "Mathematical Models in Immunology." Nauka, Moscow, 1980.

140. Martynuk, D. I., "Lectures on the Theory of Stability of Solutions of Systems with Retardations." Inst. Mat. Acad. Nauk Uk. SSR, Kiev, 1971.
141. "Mathematical Methodes in clinical practice." Nauka, Novosibirsk, 1978.
142. Matrosov, V. M., About stability of motion. *Appl. Math. Mech.* **26**(6), 992–1002 (1962).
143. Mazurov, V. M., Malov, D. I. and Salomykov, V. I. Systems of control of PH value in absorbtional column with recycle. *Himich. Promysh.* (4), 63–65 (1974).
144. Melkumjan, D. O., "Analysis of System by the Method of Logarithmic Derivative." Energoatomizdat, Moscow, 1981.
145. Melvin, W. R., Liapunov's direct method applied to neutral functional differential equations. *J. Math. Anal. Appl.* **49**(1), 47–58 (1975).
146. Michelkevich, V. N. and Chabanov, Ju. A., Identification and synthesis of system of control of process of infeed grinding. *Electromashinostroenie i Electrooborudovanie*, (*Kiev*). **31**, 42–42 (1980).
147. Migulin, V. V., Medvedev, V. I., Mustel', E. R. and Parygin, V. M., "Foundations of Oscillation Theory, Moscow, Nauka, 1978.
148. Minkin, S. I. and Skljarov, Ju. S., Analysis of transient in long line of direct current by the methods of theory of differential difference equations. *Izv. Vyss. Ucebn. Zaved. Elektromehanika* (7), 687–694 (1975).
149. Minorsky, N. (1) Control Problems. *J. Franklin Inst.* **232**(6), 519–551 (1941). (2) Self-excited oscillations in dynamical systems possessing retarded actions. *J. Appl. Mech.* (9), 65–71 (1942).
150. Misnik, A. F., Liapunov direct method for neutral equations. *Trudy Sem. Teor. Differencial. Uravnenii s Otklon. Argumentom Univ. Druzby Narodov* **6**, *Patrisa Lumumby* 78–108 (1968).
151. Misnik, A. F. and Nosov, V. R., About stability of difference-differential equations of neutral type. *Sbornik Nauchnyh Rabot Aspirantov Univ. Lumumba* **1**, 43–55 (1968).
152. Mitropolskii, Ju. A. and Filchakov, P. F., About solutions of nonlinear differential equations with deviating argument with the aid of series. *Dokl. AN USSR* **212**(5), 1059–1062 (1973).
153. Mitropolskii, Ju. A. and Martynuk, D. I., "Periodic and Quasi-Periodic Oscillations in Systems with Delay." Vysha Shkola, Kiev, 1979.
154. Mjasnikov, N. N., Theory of direct Vyshnegradskii control and influence of delay. *Izv. AN USSR, OTN*(9), 1217–1228 (1953).
155. Mladov, A. G., "Systems of Differential Equations and Liapunov Stability." Vysshai Shkola, Moscow, 1966.
156. Myshkis, A. D. (1) General theory of differential equations with delay. *Uspehi, Mat. Nauk* **4**(5), 99–141 (1949). [*Engl. Transl. AMS* **55**, 1–62 (1951), and *Ser. 1*, **4**, 207–267 (1962).] (2) "General Theorems of Theory of Ordinary Differential Equations in Nonclassical Cases." 1 Summer Mathematical School, pp. 45–116. Nauka Dumka, Kiev, 1964. (3) "Linear Differential Equations with Delaying Argument." Nauka, Moscow, 1972. (4) Some problems in the theory of differential equations with deviating arguments. *Amer. Math. Soc. Trans.* **2**(105), 237–246 (1976). (5) On some problems of the theory of differential equations with deviating argument. *Uspehi. Mat. Nauk.* **32**(2), 173–202 (1977).
157. Myshkis, A. D. and El'sgol'tz, L. E., The status and problems of the theory of differential equations with deviating argument. *Uspehi, Mat. Nauk.* **22**(2), 21–57 (1967).
158. Myshkis, A. D., Shimanov, S. N. and El'sgol'tz L. E., Stability and oscillations of systems with time-lag. *Trudy of ICNO-1, Kiev* **2**, 241–267 (1963).
159. Nechaev, Ju. N. and Fedorov, R. M., "Theory of aircraft gas-turbine engines." Mashinostroenie, Moscow, 1977.
160. Neimark, Ju. I. (1) D-subdivision and spaces of quasi-polynomials. *Prikl. Mat. Meh.* **13**(4), 349–380 (1949). (2) "Dynamical Systems and Controlled Processes." Nauka, Moscow, 1978.

161. Netushil, A. V., Plutes, V. S. and Vlasov, Ju. A., To the problem of application of SAR with compensation of delays and varying of the object parameters. *Izv. Vyss. Ucebn. Zaved. Electromehanika* (8), 882–891 (1976).
162. Nosov, V. R. (1) On a problem arising in the theory of optimal control with aftereffect. *Prikl. Mat. Meh.* **2**, 399–403 (1966). (2) Linear boundary-value problems with small lags. *Differencial'nye Uravnenija* **3**, 1025–1028 (1967). (3) Comparison theorems for scalar retarded equations. *Trudy Sem. Teor. Differencial. Uravnenii s Otklon. Argumentom Druzby Narodov Patrisa Lumumby* **4**, 247–253 (1967). (4) Sur l'arternative de Fredholm pour les systèmes linéaires généraux a coefficients periodiques et à argument deplacé. *C. R. Acad. Sci.* **270**(17), 1097–1100 (1970). (5) Solutions periodiques des systémes differentiels fonctionnels quasilinéaires. *C.R. Acad. Sci.* **270**(18), 1170–1173 (1970). (6) Periodic solutions of linear equations of general form with deviating arguments. *Differencial'nye Uravnenya* **7**, 639–650 (1971). (7) On existence of periodic solutions of linear systems of general form with distributed deviations. *Differencial'nye Uravnenija* **7**, 2168–2175 (1971). (8) On comparison theorems. *Trudy Sem. Teor. Differencial. Uravnenii s Otklon. Argumentum Univ. Druzby Narodov Patrisa Lumumby* **8**, 143–153 (1972). (9) Periodic solutions of quasilinear functional differential equations. *Izv. Vyss. Ucebn. Zaved. Matematica* **5**, 55–62 (1973). (10) Matrosov criterion of stability for retarded functional differential equations. *Differencial'nye Uravnenija* **7**, 1202–1207 (1977). (11) Investigations of stability and quality of transient process in model reference adaptive systems. *VII Internationale Konferenz uber Nichtlineare Schwingungen, Band II*, **2**, 173–182 (1977). (12) Identification of linear and some nonlinear systems with the aid of adaptive models. *In* "Optimization of Dynamic Systems," 110–114. Minsk, 1978. (13) Perron's theorem for autonomous and periodic systems of functional differential equations. *Trudy Sem. Teor. Differencial. Uravnenii s Otklon. Argumentom Univ. Druzby Narodov Patrisa Lumumby*. 44–51 (1979).
163. Nosov, V. R. and Prokopov, B. I. (1) Estimation of transient process in model reference adaptive systems. *Izv. AN USSR, Techn. Kybernetika*. **4**, 186–191 (1976). (2) On asymptotic stability of adaptive scalar retarded systems. *Differencial'nye Uravnenija* **13**, 1528–1531 (1977). (3) Global asymptotic stability of model reference adaptive systems. *Prikl. Mat. Meh.* **41**, 850–858 (1977).
164. Nosov, V. R. and Furasov, V. D., Stability of discrete systems relative to given variables and convergence of some optimization algorithms. *J. Vycis. Mat. Phys.* **19**, 316–328 (1979).
165. Nyquist, H., Regeneration theory. *Bell Syst. Tech. J.* **11**(1), 126–147 (1932).
166. Osipov, IU. S., Stabilization of nonlinear control systems with delay in critical case of one zero root. *Differencial'nye Uravnenija* **1**(7), 908–922 (1965).
167. Pak, V. E. and Prokopjev, V. P., About stability of systems with aftereffect in critical case of four zero roots. *Math. Zapiski Ural. Univ.* **7**(4), 76–82 (1970).
168. Parks, P. C. Liapunov redesign of model reference adaptive control systems. *IEEE Trans. Automat. Control* **11**(3), 362–367.
169. Pazdera, J. S. and Pottinger, H. J., Linear system identification via Liapunov design techniques. (Joint Automat. Control Conference, 10th Univ. Colorado, New York, ASME, 795–801, 1969.
170. Pinney, E., "Ordinary difference-differential equations," Univ. of California Press, Berkeley, 1958.
171. Pisarenko, V. G. (1) "Problems of Relativistic Dynamics of Many Bodies and Nonlinear Theory of Field. Nauka Dumka, Kiev, 1974. (2) Equations with deviating argument in the problem of many gravitated electricity charged bodies with delays of the forces of interaction. *In* "Differential Equations with deviating argument," pp. 255–269. Nauka Dumka, Kiev, 1977.
172. Pjatnitzkii, E. S., Structural stability of single-circuit systems with delay. *Automat. Remote Control* **23**(7), 852–862 (1962).

173. Pletnev, G. P., "Automatic control and protection of heatenergetic designs of electric stations." Energia, Moscow, 1970.
174. Political and related models *In* "Modeles in applied mathematics," Vol. 2 (W. F. Lucas, ed.) Springer-Verlag, Berlin and New York, 1983.
175. Pontryagin, L. S., On the zeros of some elementary transcendental functions. *Izv. Akad. Nauk SSSR* **6**(3), 115–134 (1942).
176. Popov, V. M. and Halanay, A., About stability of nonlinear controlled systems with delay. *Automat. Remote Control*, **23**(7), 849–851 (1962).
177. Postnikov, M. M., "Stable Polynomials." Nauka, Moscow, 1982.
178. Prokopjev, V. P., About stability of systems with delay in critical case of zero roots. *Izv. Vusov, Math.* (1), 88–94 (1967).
179. Rao, D. and Ramakrishna, Neutral functional differential equations and extensions of Liapunov's method. *Internat. J. Control* **10**(4), 369–375 (1969).
180. Razumikhin, B. S. (1) On the stability of systems with a delay. *Prikl. Mat. Meh.* **20**(4), 500–512 (1956). (2) Application of Liapunov method to problems in the stability of systems with a delay. *Automat. i Telemeh.* (21), 740–749 (1960).
181. Razvan, V., "Absolute Stability of Automatic Systems with Delay." Nauka, Moscow, 1983.
182. Real, J., Stochastic partial differential equations with delay. *Stochastics* **8**(2), 81–102 (1982).
183. Repin, Yu. M. (1) Stability conditions of systems of linear differential equations for any delays. *Proc. Ural Univ.* **23**(2), 34–41 (1960). (2) Quadratic Liapunov functionals for systems with delay. *Prikl. Mat. Meh.* **29**, 564–566 (1965).
184. Reswick, J. B., A Delay-Line Controller. Trudy First IFAC Congress, Moscow, 1960.
185. Richaud, L., La régulation P.I.D. des processus retardés. II. Systèmes du second ordre. *Automatisme* **15**(12), 626–638 (1970).
186. Riesz, F., Sz. and Nagy, B., "Leçons d'analyse fonctionnelle." Akademiai Kiado, 1972.
187. Rjabov, Ju. A. (1) Some asymptotic properties of linear systems with a small delay. *Trudy Sem. Teor. Differencial. Uravnenii s Otklon. Argumentom Univ. Druzby Narodov Patrisa Lumumby* **3**, 153–164 (1965). (2) On approximation of solutions of nonlinear retarded differential equations. *Trudy Sem. Teor. Differencial. Uravnenii s Otklon. Argumentom Univ. Druzby Naradov Patrisa Lumumby* **3**, 165–185 (1965).
188. Rohella, R. S. and Chatterjee, B., Effect of time delay on nonlinear systems. *J. Inst. Electron. Telecomm. Eng. (New Delhi)* **25**(9), 386–388 (1979).
189. Rotach, V. Ja., "Dynamics of Industrial Automatic Control Systems. Energia, Moscow, 1973.
190. Rotach, V. Ja and Stafeichuk, B. G., Application of linear Smith's predictor for control of systems with delay. *Energetica* (2) (1966).
191. Rouche, N. Abetz, P. and Laloy, M., "Liapunov's Direct Method in Stability Theory." MIR, Moscow, 1980.
192. Rouche, N. and Mawhin, J., "Equations Differentielles Ordinaires," Vol. 2. Mason, Paris, 1973.
193. Rubanik, V. P., "Oscillations of Quasilinear Systems with Retardations." Nauka, Moscow, 1969.
194. Rutman, M. A., About bounded solutions of linear differential and differential-difference equations. *Tr. Odess. Gidromet. Inst.* (20), 3–7 (1959).
195. Sawano, K., Exponential asymptotic stability for functional differential equations with infinite retardations. *Tohôku Math. J.* **31**(3), 363–382 (1979).
196. Schipanov, G. V., Theory and methods of design of automatic controllers. *Automat. Remote Control* (1), 49–56 (1939).
197. Seifert, G., Liapunov–Razumikhin conditions for stability and boundedness of functional differential equations of Volterra type. *J. Differential Equations* **14**(3), 424–430 (1973).

198. Serebrjakova, I. V., Methods of solutions of differential equations with deviating argument in 18th and 19th centuries. *Lumumby Trudy Sem. Teor. Differencial. Uravnenii s Otklon Argumentom Univ. Druzby Narodov Patrisa* **10**, 41–68 (1977).
199. Shaichet, L. E. (1) Investigation of stability of stochastic systems with delay by the method of Liapunov functionals. *Problems of Information Transmission* **11**(4), 70–76 (1975). (2) Stability in the first approximation of stochastic retarded systems. *Appl. Math. Mech.* **40**(6), 1116–1121 (1976). (3) On the optimal control of one class of stochastic differential equations with partial derivatives. *Mat. Zametki* **31**(6), 933–936 (1982).
200. Sharkovskii, A. N. and Romanenko, E. Ju., "Asymptotical Behavior of Solutions of Differential-Difference Equations," pp. 171–200. Institute of Math. of Acad. of Sci. of Ukr. SSR, Kiev, 1981.
201. Sheridan, T. B. and Ferrell, W. R., "Man–Machine Systems: Information, Control and Decision Models of Human Performance." MIT Press, Cambridge, Massachusetts, 1974.
202. Shevyakov, A. A. (ed.), "Control Theory of Power Plants of Flying Vehicles." Mashinostroenie, Moscow, 1976.
203. Shigin, E. K. Classification of dynamical models of control objects of chemical technological processes. *Automat. Remote Control* (6), 145–162 (1968).
204. Shil'man, S. V., "Method of Generating Functions in the Theory of Dynamical Systems." Nauka, Moscow, 1978.
205. Shimanov, S. N. (1) Instability of motion of a system with time lag. *Prikl. Mat. Meh.* **24**(1), 55–63 (1960). (2) On stability in the critical case of a zero root for delayed systems. *Prikl. Mat. Mech.* **24**(3), 447–457 (1960). (3) To the theory of retarded linear periodic differential equations. *Prikl. Mat. Meh.* **27**(3), 450–458 (1963).
206. Shimbell, A., Contributions to the mathematical biophysics of the central nervous system with special reference to learning. *Bull. Math. Biophys.* (12), 241–275 (1950).
207. Shui-Nee Chow and Hale, J., "Methods of Bifurcation Theory," Vol. 251. Springer-Verlag, Berlin and New York, 1982.
208. Sievert, R. M., Zur theoretisch-mathematischen Behandlung des Problems des biologischen Strahlenwirkung. *Acta Radiologica* **22**, 237–251 (1941).
209. Silkowski, R., A star-shaped condition for stability of linear retarded functional differential equations. *Proc. Roy. Soc. Edinburgh Sect. A* **83**, 189–198 (1979).
210. Sinha, A. S. C., On stability of solutions of some third and fourth order delay-differential equations. *Information and Control* **23**(2), 165–172 (1973).
211. Slemrod, M. and Infante, E. F., Asymptotic stability criteria for linear systems of difference-differential equations of neutral type and their discrete analogues. *J. Math. Anal. Appl.* **38**(2), 399–415 (1972).
212. Slusarchuk, V. E., Strong absolutely asymptotically stable solutions of linear retarded differential equations in Banach spaces. *Differencial'nye Uravnenija* **15**(9), 1614–1619 (1979).
213. Smirnov, A. I., "Aeroelastic Stability of Flying Vehicles," Mashinostroenie, Moscow, 1980.
214. Smith, O. J. M. (1) "Feedback Control Systems." MacGraw-Hill, New York, 1958. (2) A Controller to overcome dead time. *JSA Journal*, **6**(2) (1959).
215. Smol'nikov, L. P. and Moskvin, V. M., Design of controllers with Smith predictor for plants with delay. *Izv. Leningrad Electrotechn. Ins.* **111**, 74–81 (1972).
216. Snow, W., Existence, uniqueness and stability for nonlinear differential-difference equations in the neutral case. Preprint, New York University, New York, 1965.
217. Sokolov, A. A. Stability criterion of linear controlled systems with distributed parameters and its applications. *Engin. Sbornik* **2**(2), 3–26 (1946).
218. Solodov, A. V. and Solodova, E. A., "Systems with Variable Delay." Nauka, Moscow, 1980.
219. Solodovnikov, V. A. (1) Use of operator method in studying of the speed control of

hydroturbines. *Avtomat. i Telemeh.* (1), 5–20 (1941). (2) "Theory of automatic control," Vol. 1. Mashinostrienie, Moscow, 1967.
220. Solodovnikov, V. V., Filimonov, A. B. Design of controllers for plant with delay. *Izv. AN USSR, Tech. Kybernet.* (1), 168–177 (1979).
221. Starik. L. K., Coupled quasilinear oscillating systems with a source of energy involving retarded connections. *Trudy Sem. Teor. Differencial. Uravnenii s Otklon. Argumentom. Univ. Druzby Narodov Patrisa Lumumby* 3, 119–132 (1965).
222. Stech, H. W., The effect of time lags on the stability of the equilibrium state of a population growth equation. *J. Math. Biol.* 5(2), 115–120 (1978).
223. Stokes, A. (1) A Floquet theory for functional differential equations. *Proc. Nat. Acad. Sci. USA* 48(8), 1330–1334 (1962). (2) Stability of functional differential equations with perturbed lags. *J. Math. Anal. Appl.* 47(3), 604–619 (1974).
224. Svirejev, Ju. P. and Logofet, D. O., "Stability of Biological Populations." Nauka, Moscow, 1978.
225. The damping effect of time lag (art. editorial). *Engineer* 163(5) (1937).
226. *Trudy Sem. Teor. Differencial. Uravnenii s Otklon. Argumentom Univ. Druzby Narodov Patrisa Lumumby* 1 (1962); 2 (1963); 3 (1965); 4 (1967); 5 (1967); 6 (1968); 7 (1969); 8 (1972); 9 (1975); 10 (1977); 11 (1979).
227. Tychonov, A. N., Sur les équations fonctionelles de Voltera et leurs applications à certains problèmes de la physique mathématique. *Bull. de l'Univ. d'Etat Moscou. Sect. A* (1), 1–25 (1938).
228. Tyshkevich, V. A., "Some Problems of Stability of Functional Differential Equations." Nauka Dumka, Kiev, 1981.
229. Tzaljuk, Z. B. Volterra integral equations. *Naucn.-Techn. Informacija (VINITI)* 15, 131–198 (1977).
230. Tzar'kov, E. F. (1) Asymptotic exponential mean-square stability of the trivial solution of stochastic functional-differential equations. *Theory Probab.* (4), 871–875 (1976). (2) Exponential p-stability of trivial solution of functional-differential equations. *Theory Probab. Appl.* (2), 445–448 (1978).
231. Tzar'kov, E. F. and Engel'son, L. E. (1) Statistical solutions of linear systems of functional differential equations. *In* "Topological spaces and mappings," pp. 144–153. Riga University, 1981. (2) Liapunov functional for linear periodic differential equations with aftereffects. *In* "Topological spaces and mapping," pp. 117–136. Riga University, 1983.
232. Tzypkin, Ja. Z. (1) Stability of system with delayed feedback. *Automat. Remote Control* 7(2), 107–129 (1946). (2) The power of stability of delayed feedback systems *Automat. Remote Control* 8(3), 145–155 (1947). (3) Stability of a class of controlled systems with distributed parameters. *Automat. Remote Control* 9(3), 176–189 (1948).
233. Valeev, K. G., Linear differential equations with sinusoidal coefficients and constant retardations. *Proc. ICNO-1* 2, 100–119 (1963).
234. Vasiljeva, A. B. and Butusov, V. F., "Asymptotic Expansions of Solutions of Singular Perturbed Equations." Nauka, Moscow, 1973.
235. Vasiljeva, A. B. and Rodionov, A. M., Applications of the method of perturbations to retarded equations with small delays. *Trudy Sem. Teor. Differencial. Uravnenni s Otklon. Argumentom Univ. Druzby Narodov Patrisa Lumumby* 1, 20–27(1962),
236. Velarde, M. C., Dissipative structures and oscillation in reaction-diffusion models with and without time-delay. *In* "Stability of Thermodynamics Systems," (G. Lebon, ed.), Springer-Verlag, Berlin and New York, 1982.
237. Venkatesh, Y. V., "Energy Methods in Time-Varying System Stability and Instability Analyses." Springer-Verlag, Berlin and New York, 1977.
238. Volkov, V. Ja. and Kuprijanov, N. S., Stability of linear systems with many delays. *Technicçeskaya Kibernetika* (5), 170–175 (1968).

239. Volterra, V. (1) Sulle equazioni integrodifferenziali della teorie dell'elasticita. *Atti Accad. Lincei* (18), 295 (1909). (2) Sur la theorie mathematique des phenomenes hereditaires. *J. Math. Pures Appl.* (7), 249–298 (1928). (3) "Theorie mathematique de la lutte pour la vie." Gauthier-Villars, Paris, 1931.
240. Vorobjev, Ju. V., "Investigations of Control of Steam Turbine." Gosenergoizdat, Moscow, 1950.
241. Voronov, A. A. (1) "Stability, Controllability and Observability." Nauka, Moscow, 1979. (2) "Foundation of the Theory of Automatic Regulation." Energoizdat, Moscow, 1981.
242. Vostrikov, A. S. and Utkin, V. I., Franzusova, G. A., Systems with derivative of phase-vector in control. *Automat. Remote Control* (3), 22–25 (1982).
243. Voznesenskii, I. N., To choice of control of power-and-heat supply turbine. *Soviet Energooborudovanie (Moscow)* (1934).
244. Walker, J. A., "Dynamical Systems and Evolution Equations: Theory and Applications." Plenum, New York, 1980.
245. Wangersky, P. J. and Cunningham, W. J., Time lag in prey-predation population models. *Ecology* **38**, 136–139 (1957).
246. Worz-Busekros, A., Global stability in ecological systems with continuous time delay. *SIAM J. Appl. Math.* **55**(1), 123–134 (1978).
247. Wright, E. M. (1) Linear differential difference equations. *Proc. Cambridge Philos. Soc.* **44**, 179–185 (1948). (2) The linear difference-differential equation with asymptotically constant coefficients. *Amer. J. Math.* **70**(2), 221–238 (1948). (3) The stability of solutions of non-linear difference-differential equations. *Proc. Roy. Soc. Edinburgh, Sect. A* **63**(1), 18–26 (1950).
248. Yorke, J. (1) Some extensions of Liapunov's second method. *SIAM Diff. Integ. Equat. Pa.*, 206–207 (1969). (2) Asymptotic stability for one dimensional differential-delay equations. *J. Differential Equations* **7**(1), 189–202 (1970).
249. Yoshizawa, T., "Stability Theory and the Existence of Periodic Solutions and Almost Periodic Solutions." Springer-Verlag, Berlin and New York, 1975.
250. Zaharov, A. A., Kolesov, Ju. S., Spokojnov, A. N. and Fedotov, N. B. Theoretical explanation of ten years oscillation cycles of quantities of animals in Canada and Iakoutia. *In* "Studies in Stability and Oscillations," pp. 82–131. Iaroslavl, 1982.
251. Zhadanov, V. I. On the one-dimensional symmetric two-body problem of classical electrodynamics.—*Internat. J. Theoret. Phys.* **15**, 157–167 (1976).
252. Zubov, V. I. (1) "Mathematical Methods of Investigation of Controlled Systems." Sudpromgiz, 1959. (2) Stochastic behavior of system with finite number of phase states and after-effect. *Differencial 'nye Uravnenija* **15**(3), 387–391 (1979). (3) "Analytical Dynamics of Systems of Bodies." Leningrad Univ., Leningrad, 1983.
253. Zverkin, A. M. (1) Dependence of the stability of solutions of linear differential equations with lag upon the choice of the initial moment. *Vestnik Moscov Univ. Ser. I. Mat. Meh. Astr. Him.* **5**, 15–20 (1959). (2) Theorems of existence and uniqueness for equations with deviating argument in critical case. Trudy Sem. Teor. *Differencial. Uravnenii s. Otklon. Argumentom Univ. Druzby Naradov Patrisa Lumumby* **1**, 37–46 (1962). (3) Differential-difference equations with periodic coefficients.—Appendix to Russian translation of "Differential Difference Equations" (Bellman and Cooke). MIR, Moscow, 1967. (4) Expansions of solutions of differential difference equations in series. *Trudy Sem. Teor. Differencial. Uravenii s Otklon. Argumentom Univ. Druzby Narodov Patrisa Lumumby* **4**, 3–50 (1967). (5) Differential equations with deviating argument. *Trudy 5 Letney Mat. Shkoly*, 307–399, Nauka Dumka, Kiev, (1968).

Index

A

Aeroautoelasticity, 120
Antenna, 67

B

Boundedness, 159
Buss, 124

C

Condition
  Perron, 159
  self-excitation, 150
  splicing, 22
Controller
  adaptive, 43, 96
  P, 35, 37, 38, 81
  PD, 37, 118, 132
  PID, 37, 118
  Reswick, 41
  Smith, 42
Criterion
  frequency, 60
  integral, 63, 70
  Matrosov, 77
  Michailov, 60, 119, 123
  Nyquist, 61, 62, 119
  Popov–Halanay, 107
  Tzypkin, 65

D

Delay
  adjustment, 100
  bounded, 19, 72, 130, 148
  random, 170
  small, 27
  unbounded, 19, 21, 74
Diagram, Vyshnegradskii, 55, 58, 117, 118
Dynamics, relativistic, 13

E

Equation
  differential difference, 19, 22, 34, 53
  for moments, 166
  linear, 25
    autonomous, 31, 48, 113, 196
    periodic, 108, 160
    scalar, 83, 134, 199
    second-order, 85, 180
Estimate, *a priori,* 24, 26

F

FDE, 36
Flutter, 125
Formula, Itô, 166, 169, 172, 177, 180
Function
  almost-periodic, 129
  characteristic, 31, 33, 50, 69, 114
  transfer, 1, 35, 37, 69
  unsteady unit step, 121
Functional
  degenerated, 132, 140, 152
  Liapunov, 126, 156, 159
    –Krasovskii, 72, 80, 89

H

Hodograph, Michailov, 61, 62, 119, 120

I

Ideal predictor, 36
Inequality
  Chebyshev, 189
  functional, 127, 129, 152, 194
Instability, 42, 45, 142, 154

L

Lag
  informational, 4
  technological, 1
  transportation, 3
Laplace transform, 49, 114

M

Method
  D subdivision, 55, 117
  Liapunov direct, 71, 103, 126, 167, 180, 187
  steps, 19, 48, 111, 154, 161, 167
Model
  cutting, 8
  immune response, 15
  infeed grinding, 8
  linear stochastic aeroautoelastic, 123
  population dynamics, 13
  predator–prey, 15
  virus disease, 90
Multiplier, Floquet, 108, 161

N

NFDE, 10, 19, 21
  linear, 27
    autonomous, 33, 113
    with bounded delay, 130, 148

O

Oscillator, 9

P

Problem, initial value, 21, 23, 164
Process
  Markov, 166, 194
  transient, 38
  Wiener, 164

Q

Quasi-polynomial, 32, 53

R

Reactor, 1, 81, 132
  nuclear, 8, 87
Response, frequency, 61
RFDE, 19
  linear, 25, 48
  with bounded delay, 19, 72
  with unbounded delay, 19, 21, 74

S

Sector, 145, 151
SFDE, 123, 164
SNFDE, 193
Solution
  ergodic, 190
  f bounded, 128, 130
  Floquet, 110, 161
  periodic, 189
  representation, 29, 33
  stationary, 187
Space
  $C[-h,0]$, 19
  $C[-\infty,0]$, 18, 21
  $CB[-\infty, 0]$, 19, 21, 165
  $L_\infty[-h,0]$, 22
  $L_\infty[-\infty,0]$, 24
SRFDE, 6, 164
Stability, 44, 113, 164
  absolute, 106
  domain, 117, 124, 180, 199
  exponential, 79, 170, 188, 192
  global, 14, 128, 132, 136, 140
  in the first approximation, 75, 191
  invariant set, 152
  Liapunov, 44
  linear autonomous NFDE, 113
  linear autonomous RFDE, 48
  under steady acting disturbances, 75, 79, 99, 101

Stabilization, ship, 4
Stable, 44, 45, 73, 127, 157
  asymptotically, 46, 51, 73, 115, 128
    uniformly, 47, 73, 153
  f, 127, 129, 130, 152
    asymptotically, 127, 129, 130, 152
  L_2, 49, 50, 114
  mean square, 168
    asymptotically, 168, 171, 173, 176, 194, 196
  p, 168
    asymptotically, 168, 169, 193
    with respect to probability, 168, 170, 193
System
  central nervous, 17, 143
  distributed self-oscillatory, 150
  feedback, 35
    single-loop, 35, 38
    two-loop, 39
  man–machine, 5
  with lossless transmission line, 10

T

Theorem
  Barbashin–Krasovskii, 76
  Chebotarev, 55, 117
  Chetaev, 142, 146, 155
  Esclangon, 159
  Krasovskii, 72
  Pontriagin, 54, 117
  Razumichin, 72
  Zubov, 33
Time lag, 2, 3
Turbojet, 6

V

Viscoelasticity, 103

U

Uniqueness, 21, 23, 166, 194

RAYMOND H. FOGLER LIBRARY
DATE DUE

**BOOKS ARE SUBJECT TO
RECALL AFTER TWO WEEKS**

NOV 2 5 1987